ALSO BY ALAN LIGHTMAN

Ghost

The Discoveries

A Sense of the Mysterious

Reunion

The Diagnosis

Dance for Two: Essays

Good Benito

Einstein's Dreams

Great Ideas in Physics

Time for the Stars

Ancient Light

Origins

Song of Two Worlds

Mr g

Mr g

A Novel About the Creation

ALAN LIGHTMAN

Pantheon Books   New York

Library of Congress Cataloging-in-Publication Data
Lightman, Alan P., [date]
Mr g : a novel about the creation / Alan Lightman.
p. cm.
ISBN 978-0-307-37999-3
1. God—Fiction. 2. Creation—Fiction. 3. Evolution—Fiction. 4. Universe—Fiction. I. Title. II. Title: Mister g : a novel about the creation.
PS3562.I45397M7 2011 813'.54—dc22 2011000095

www.pantheonbooks.com

Jacket design by Rodrigo Corral Design
Book design by Virginia Tan

Printed in the United States of America
First Edition

2 4 6 8 9 7 5 3 1

This book is dedicated to my brother
Ronnie Lightman
(1951–2010)

Mr g

Time

As I remember, I had just woken up from a nap when I decided to create the universe.

Not much was happening at that time. As a matter of fact, time didn't exist. Nor space. When you looked out into the Void, you were really looking at nothing more than your own thought. And if you tried to picture wind or stars or water, you could not give form or texture to your notions.

Those things did not exist. Smooth, rough, waxy, sharp, prickly, brittle—even qualities such as these lacked meaning. Practically everything slept in an infinite torpor of potentiality. I knew that I could make whatever I wanted. But that was the problem. Unlimited possibilities bring unlimited indecision. When I thought about this particular creation or that, uncertain about how each thing would turn out, I grew anxious and went back to sleep. But at a particular moment, I managed . . . if not exactly to sweep aside my doubts, at least to take a chance.

Almost immediately, it seemed, my aunt Penelope asked me why I would want to do such a thing. Wasn't I comfortable with the emptiness just as it was? Yes, yes, I said, of course, but . . . You could mess things up, said my aunt. Leave Him alone, said Uncle Deva. Uncle toddled over and stood beside me in his dear way. Please don't tell me what to do, retorted my aunt. Then she turned and stared hard at me. Her hair, uncombed and knotted as usual, drooped down to her bulky shoulders. Well? she said, and waited. I never liked it when Aunt Penelope glowered at me. I think I'm going to do it, I finally said. It was the first decision I'd made in eons of unmeasured existence, and it felt good to have decided something. Or rather, to have decided that something had to be done, that a change was in the offing. I had chosen to replace nothingness with something. Something is not nothing. Something could be anything. My imagination reeled. From now on, there would be a future, a present, and a past. A past of nothingness, and then a future of something.

In fact, I had just created time. But unintentionally. It was just that my resolution to act, to make things, to put an end to the unceasing absence of happenings, required time. By deciding to create something, I had pressed an arrow into the shapeless and unending Void, an arrow that pointed in the direction of the future. Henceforth, there would be a before and an after, a continuing stream of successive events, a movement away from the past and towards the future—in other words, a journey through time. Time necessarily came before light and dark, matter and energy, even space. Time was my first creation.

Sometimes, the absence of a thing is not noticed until it is present. With the invention of time, events that had once merged together in one amorphous clot began to take shape. Each event could now be enveloped by a slipcover of time, separating it from all other events. Every motion or thought or the slightest happenstance could be ordered and placed exactly in time. For example, I realized that I had been sleeping for a very long time. And near me—but I couldn't say how near, because I had not yet created space—Aunt Penelope and Uncle Deva had also been sleeping, their loud snores rising and falling like something or other, their tossings and turnings unfolding in time. And their interminable bickering could now be identified with moments of wakefulness, which in turn could be understood as taking place between periods of sleep. I refused to think how much time I had wasted. In fact, we had all slept in a kind of pleasant amnesia, a swoon, an infinite senselessness. In various ways, had we not luxuriated in the unstructured Void, unaccountable for our actions? Yes, unaccountable. Because without time, there could be no reactions to actions, no consequences. Without time, decisions need not be considered for their implications and effects. We had all been drifting in a comfortable Void without responsibilities.

See, my aunt complained when it became apparent that we were now conscious of time. I told you that you would mess things up. She shot Uncle a look of disapproval, as if he had encouraged me to act as I had, and then she began an unhappy summary of the various things that she had done and not done during the immediate past, then during the past before that, and so on, back and back through the now visible chasms of time, until Uncle begged her to stop. You should never have created the past and the future, she said. We were happy here. See, now I must say *were*, when before . . . Oh! There it is again. It was nicer when everything happened at once. I can't stand to think about the future. But don't you think that we have some responsibility to the future? I suggested. To all the things and beings I might create? Nonsense, shrieked Aunt Penelope. What a foolish argument. You have no responsibility to things that don't yet exist and won't ever exist if you could just keep your big thoughts to yourself. But it's too late now, she went on. I can feel time. I can feel the future. She had gotten herself into one of her states, and the Void twisted and throbbed with her displeasure.

Mr g

Gently, Uncle caressed her. For the first time ever, she responded to his touch. Her ranting diminished. Soon after, she realized that her hair needed combing, and that was the beginning of something and probably all for the best.

The Void

Time trickled by for certain periods and intervals. At other moments, it gushed ahead, flew headlong into the future, then braked and slowed again to a trickle. Having created time, I had not decided whether it should flow uniformly or in fits and starts. But the matter was more nettlesome than that. Since I had not yet created clocks, one could not say what constituted a smooth versus a choppy passage of time. There was nothing to measure it. Perhaps the movement of time might even be relative to the observer. Or perhaps it could be only perception. At the beginning, all any of us knew for sure was that time flowed. I didn't feel like committing myself right away to one possibility or the other—I had been pondering much as it was—so I decided to decide the texture of time at a future date.

Whether smooth or choppy, the creation of time had already altered the Void. Before time, we did not move through the Void so much as we experienced it all at once. Or rather, the Void clung to our beings, the Void contained our thoughts, the Void constituted the nothingness against which our somethingness existed. After time, the Void remained an infinite and unchanging emptiness, but now one could travel through it as well as think it, one could say that one had been at a particular place of the Void at one moment and another place at a later moment. Not that the Void had signposts or markers to designate definite locations—the Void was perfectly smooth, empty, and without any shape—but we understood that such locations existed in principle, and we could pass from one to the other over a period of time. And even though the Void was empty, totally empty, at various moments we could glimpse the faintest of features—wispy draperies, veils, gossamer ridges, valleys of nothingness—that would briefly appear and then disappear. Such elusive structures arose between the seams of the many layers of nothingness packed on top of one another, where they did not fit precisely together. If one began moving towards a particular of these evanescent topographies, it would vanish in short order but nevertheless provide a momentary route of travel, a fleeting destination, a temporary break from the complete formlessness of the Void.

I spent great swaths of time moving through the

Void. Although empty, the Void constantly beckoned with its infinite wells of possibilities. I would travel in a certain direction for a long time, moving through vapors of nothingness, then suddenly decide I wanted to explore new territory and would turn to the right or to the left and travel for a long period in another direction. Occasionally, I would do an about-face and go back the way I had come, traveling for extremely long durations through one empty place and then another and another. Frequently, I had no particular destination in mind but was merely following a natural curiosity to understand how the Void had been transformed by time. Sometimes, I would play games with myself, pretending to be lost, and I would identify my position not by my innate knowledge but by estimating the amount of time passed in various directions and performing geometrical calculations. I once moved around in a spiral of ever-increasing diameters, passing near places I had passed before, virtually the same emptiness but with the most subtle changes in each repetition, minute alterations in the vacuum brought about by the lapsing of time. Sometimes I would stop altogether and just admire the quiet beauty of the Void, the serenity, the unending pilasters and balustrades of nothingness. I could never gauge the actual distance traversed in any of these outings, since space didn't exist, but I knew that vast amounts of time had transpired. At various moments, Aunt

Mr g

or Uncle would appear from behind a billowing veil of the Void, we would register surprise at having encountered each other, say hello, and go our separate ways. Such chance meetings, requiring a before and an after, never happened before the creation of time.

I can say that my long treks through the Void were pleasing. I liked being in motion, going from one place to another. With movement, I felt a heightened intensity of existence and being. And the emptiness had a way of accumulating as I traveled, the clouds and vapors of nothingness sticking to me in increasing numbers, so that I had the sensation of being cloaked by an ever-thickening garment of soft cushions. I certainly had an utter vacuum in which to think. Given that the Void was total and complete emptiness, I proceeded to fill it with my thoughts, and those thoughts served as signposts of a kind. Here is where I had the idea of the universal ratio of circumference to diameter, the number π. Over there is where I had the notion of a spectrum of colors. And so forth. The Void served as a gracious receptacle for my thoughts. It was my playground of ideas.

Then there was music. The Void had always vibrated with the music of my thoughts, but before the existence of time the totality of sounds occurred simultaneously, as if a thousand thousand notes were played all at once. Now we could hear one note following another, cascades of sound, arpeggios and glissades.

We could hear melodies. We could hear rhythms and metrical phrases gathering up time in lovely folds of sound. Duples and triples and offbeat syncopations. As we moved through the Void, all of us—Aunt Penelope, Uncle Deva, and I—were transfixed by the most exquisite sounds, the tender and melodic and rapturous oscillations of the Void.

Much of the music I devised from a scale with a fixed ratio of frequencies, generally $2^{1/12}$, since exponential powers of that number came closest to ratios of small integers like 3:2 and 4:3. Chords based on these scales were pleasing to hear. But I also experimented with quarter-tone ratios, nonharmonic ratios, and even scales with variable ratios, and these also produced beautiful music as long as two different notes were not sounded together. By varying the intensity of harmonics of each tone, I created an infinite variety of sounds.

In every place and in every moment, we were wrapped and engulfed in music. At times, the music poured forth in fierce heaving swells. At other times, it advanced in the softest little steps, delicate as a fleeting veil in the Void. Music clung to our beings as parcels of emptiness had in the past. Music went inside us. I had created music, but now music created; it lifted and remade and formed a completeness of being.

Space

I had in mind a great number of things I wanted to make. But with no previous experience with materiality, I could think of these things only in terms of their functions or qualities: the quantification of time, communication, light, shelter, et cetera. Soon I grew tired of abstractions. I wanted to touch and to feel. After all, I had been sleeping for a very long time. And I might add that I needed something new to interest me, a challenge, perhaps even other beings to surprise and amuse me. My ideas, for both animate and inanimate invention, required material existence, extension, volume. And for that I needed to create space.

Space did not appear all at once, but in a languorous progression, gradually increasing in length, width, and breadth. (I had toyed with various numbers of dimensions. Two seemed unnecessarily confining, suffocating in fact, while four or more struck me as extravagant and could lead to the misplacing of small objects. I decided my first try should be three.) As I recall, space first appeared in a minuscule round bubble that sat quietly in my mind. Then it stretched slightly in length, humming at a high pitch as it did so. For a time, the universe was a tiny ellipsoid. Slowly, breadth and width began to catch up with length, making an impatient, clucking sound. Sphericity was restored. Then, with a sigh and a low rumble, all three dimensions began to unravel at once, tumbling and sprawling into the Void.

My universe had come into being! It was tiny at first, but beautiful, a lovely little sphere. Its surfaces were smooth and silky, yet infinitely strong. It glistened. It spun slightly. And it vibrated with energy. I found that I could not create space without energy— the two were inextricably bound, as if one gave form to the other. The energy howled and struggled to break out of those smooth, silky walls, but it could not, since those walls contained all that was (except for me, Aunt, and Uncle), and it was a mathematical and tautological impossibility for anything from within to emerge without. Only the Void remained outside those walls. In its continual battle to escape the inescapable, the energy seethed and boiled at a ferocious temperature, it distorted the walls, stretched them first in one direction and then another. And then, as if in frustration, it set about stretching space itself, warping diameters and circumferences, angles and curves—contorting the very mathematics of space. The geometry, responding to the fierce stresses and strains, began to emit its own piercing hum, and the two—energy and geometry—fought with each other in a shrill screech, first the mesas and terraces of space muscling the energy by brute force, and then the energy striking back and reshaping the architecture of space. As the combat ensued, the tiny sphere that was the universe began inflating at an alarming speed.

Aunt Penelope, who in a rare moment had been quietly brushing her hair, was knocked over by the expanding sphere. Save me, she screamed to Uncle Deva, overdramatizing the situation as she often did. Uncle helped right her and steadied her. What was that thing? she shouted. The impertinence! Then, without thanking Uncle, she stomped off into the Void. Even though she had disappeared behind the folds and pleats of the vacuum, I could hear my aunt muttering: What's He done now! There's no end to this, no end. No end to this. No end. No end to this. No end . . .

Meanwhile, my universe was growing larger and larger. Once created, it seemed determined to become as fat as it could. I decided to make another. This one, I slightly pricked at the moment it came into existence,

just the smallest of flicks to see what a slight altera-
tion would bring. The little sphere began expanding
like the previous universe, but after a few moments
its expansion coasted to a halt, it briefly hovered in
a fleeting equilibrium, then it began contracting and
dwindling in size, getting smaller and smaller until
it was just the tiniest dot. Then, with a faint pop, it
disappeared altogether. I was delighted. I made other
universes. With each one, I tried a different variation.
To some, I gave a slight lateral nudge. To others, a bit
of extra spin. Some I squeezed just at the moment
of creation, to add a smidgeon of energy. In some, I
even altered the number of dimensions of space: four,
seven, sixteen, to see what might happen. And why
not try fractional dimensions, like 13.8. Some universes never came into being, unable to accommodate
all the initial conditions. Some leaped into existence
with a frightening energy and then petered out. Some
remained flaccid from the beginning; others careered
through the Void, producing high-pitched trills and
vibratos. One universe remained constant in size but
spun faster and faster until it split apart at its midsection. Several began expanding, then contracted
down to almost nothing, hesitated, and expanded
again in a kind of frothy rebirth—then repeated the
entire cycle, expansion, contraction, expansion, contraction, on and on in an unending series of births,
destructions, and rebirths.

Mr g

After a time, a gigantic number of universes were flying about—spinning on their axes, throbbing and pulsing, expanding and contracting at fantastic speed. My aunt was nowhere to be seen. Uncle Deva, as sympathetic as he was to my enterprise, had ducked for cover. In short order, as seemed almost inevitable, some of the universes began colliding with others. Each collision made a terrific explosion, sending fragments of worlds hurtling through the Void, oscillating dimensions, fractured energies.

It occurred to me that I had not carefully considered whether I should make one universe or many. Perhaps I should have been more circumspect. One universe would avoid the possibility of collisions, but then again it might become boring. One universe would have one truth. Many would have many truths. There were advantages and disadvantages to both propositions.

I sat down, centered myself, and began mulling over the matter. Then I meditated. I tried to let all thoughts flow from my mind. I breathed in the Void, breathed out the Void. Breathed in the Void, breathed out the Void. Slowly, I grew calm. A peace spread over the Void. Aunt and Uncle appeared as tiny lights, dancing together to a waltz in andante, and a peace descended on them as well, and the Void settled and sighed and drifted in unwinding time. I breathed in and I breathed out and I came to the decision that there should be only One, one universe, and the myriad temporal universes that I had made faded and dissolved, and the one universe remained.

And then, while still meditating, I decided to create quantum physics. Although I keenly appreciated the certainty of logic and clear definition, I also felt that the sharp edges of existence needed some rounding. I wanted a bit of artistic ambiguity in my creations, a measured diffusion. Perhaps quantum physics invented itself. It was gorgeous in mathematical terms. And subtle. As soon as I had created quantum physics, all objects—even though objects at that point existed only in my mind—billowed out and swelled into a haze of indefinite position. All certainties changed into probabilities, and my thoughts bifurcated into dualities: yes and no, brittle and supple, on and off. Henceforth, things could be hither and yon at the same time. The One became Many. And a great softening blanket of indeterminancy wrapped itself over the Void. My breathing slowed to a sleepy imperceptibility. Listening carefully, I could hear a billion billion tiny rattles and tinklings from all over the Void, the sound of new universes waiting to be. With the invention of quantum, each point of the Void had developed the potential to become a new universe, and that potentiality could not be denied. My creation of time, and then space, had made a universe possible—and that possibility alone, nestled within the quantum foam of the Void, was sufficient to bring into being an infinite number of universes. Soon, new universes were once again whizzing through the vacuum. I revised my earlier decision that there should be only One. Or, more precisely, my creation of quantum physics necessarily required the Many. Peering out into the Void, I tried to find my original universe, the first one I'd made. But it was hopelessly lost among billions and billions of others flying about, throbbing spheres, distended ellipsoids, gyrating cosmoses thrashing with energy. The Void trembled with rumbles and shrieks and sharp popping noises.

By and by, Aunt Penelope emerged from her hiding place, Uncle Deva from his. You've been busy, said Uncle, looking with mild annoyance at the many universes flying about. If I were you, I wouldn't get attached to any of them. You'll just be disappointed. I took Uncle's comment under advisement. Already, I was rather fond of some of the expanding spheres.

What's in those things, anyway? asked Aunt Penelope. Space, I answered. Umph, she said. Well now that we have space, I'd like, please, a chair to sit down on. I've been standing for a very long time. So I made a chair for Aunt Penelope. That chair was my first creation of matter. It had three curved legs and an octagonal back, and I'd designed it to be comfortable but not too comfortable. My aunt sat down on it without comment.

Far more awaited. I wanted to make more matter. I wanted to make galaxies and stars. I wanted to make planets. I wanted to make living creatures, and minds. But for the moment, I sat and I meditated and I gazed with contentment at the empty but vibrating universes I had made.

A Stranger Appears
in the Void

I meditated. I did meditate. I am meditating. I will meditate.

Although I had emptied my mind of thoughts, I was still conscious of the new universes flying about. I could feel the presence of the pulsating spheres, I could feel the volume and space within them. More importantly, I could feel the *potential* of space now scattered throughout the Void. While I drifted in my meditative state, I was no longer drifting through a shapeless and timeless Void, but a Void now tessellated with time and with space. The emptiness shimmered with possibilities, each tiny volume trembling with a nebulous version of everything that could possibly be, everything I might eventually create. It was a pressure, a weight, a low humming sound. And I had changed myself as well as the Void. A great *unfolding* had taken place within my being, as if every degree of consciousness had multiplied into a thousand degrees of consciousness, every possible action had branched into a thousand possible actions. With the new quantum reality, I was exquisitely aware of the fantastic number of possible decisions and possibilities at each point of existence, each with its own consequences leading to an infinite chain of potentialities. Henceforth, when I decided to create a thing, I would necessarily need to create not only that thing but every conceivable variation of the thing, each with its own probability. Existence was now multiplicity. These new sensations and realities were not unpleasant, but they did require certain adaptations and allowances.

When I finally emerged from my meditations, a stranger was standing beside me. And behind him, another creature, a fat and squat being whose countenance seemed frozen in a grin. In the unending expanse of existence, there had never been anyone other than myself, Aunt Penelope, and Uncle Deva. I was pleased to have another being to talk to, yet I was not accustomed to meeting things I had not made.

"Good day," said the stranger. "If I might take the liberty of using that expression. It will come with future creations."

"I have not invited you here," I said.

Mr g

The stranger nodded, an acknowledgment of my comment but without any apology. He was tall and thin, and he held himself both with ease and with a formality. "You have a congenial existence here," he said. "I have recently traveled through these regions, and they impart a definite tranquility. I imagine that you would want to stay here as long as possible, perhaps forever." His voice did not enter my mind in the same manner as that of Aunt and Uncle but seemed to be swept in by a breeze from the Void, even though the Void had been windless for eons of time.

"Not that I envy you," said the stranger. "But you do have comfortable circumstances."

"Too comfortable," said the grinning beast beside him.

"You forget yourself, Baphomet," said the stranger. The creature suddenly yelped, as if it had been struck a vicious blow, and then bowed three times to the tall stranger without releasing the sneer on its face.

"Pardon Baphomet," said the stranger, his gaze fastened on me. "He makes a good traveling companion." He paused. "I wonder about this emptiness," he said. "It would seem not to have any existence independent of our perception of it. An interesting substance. One could think it pleasant or unpleasant, strong or weak, and that would in fact be its reality. The mind is its own place, don't you agree? Let us take the music, for example. Quite lovely. I congratulate you. I have been listening to it and enjoying it for some time. However, is it not conceivable that to some other mind, to some other sensibility, this same music might sound . . . let us say, unlovely?"

"I, for one, do not like the music one bit," said Baphomet, and the beast quickly bowed again and grinned.

The stranger turned and stared at the beast, then turned back to me. "But there is a more serious question I wanted to ask you," he said. "Do you think it is possible for a thing and its opposite both to be true?"

Despite having been startled by the stranger and his rude companion, I found myself captivated by him, even mesmerized. I decided to answer his question.

"A thing and its opposite cannot both be true in a rational system of thought," I replied. "But rational thoughts lead only to rational thoughts, whereas irrational thoughts lead to—"

"New experiences."

"Yes," I said. "My mind encompasses both the rational and the irrational. But certain things must have logical consistency, and thus rationality."

"Exactly," said the stranger. "For example, mathematics. But logical consistency can be misleading. Even in mathematics, the truth or falsity of some theorems cannot be proven. Curious, wouldn't you say?"

"But that is beside the point. Each mathematical theorem is either true or false, whether it can be proven within the limitations of mathematics or not."

"Yes, yes," said the stranger. "I see that we can converse with each other."

As we were talking, Baphomet was doing flips and somersaults, all the while watching us with his relentless grin. His master paid no attention to him.

"Without knowing for sure," the stranger continued, "I would think that you are more fluent with the rational. It has its appeal. But the irrational permits a greater exercise of . . . shall we say, *power*. If that is your aim, of course. At the moment, you would seem to have no need to exercise your power."

"I prefer to use only the scope and magnitude of power that is required for each situation," I said. "But I have unlimited power, if necessary."

"I would very much enjoy seeing a demonstration of that sometime." The stranger moved closer. "But the target of power is more interesting than its quantity. In that regard, tell me: Would you say that the end always justifies the means? Or, in attempting to achieve your aims, do you draw the line at some degree of sacrifice and cost, beyond which you would not go?"

"I cannot consider this question in general terms."

"Ah, you do not believe in absolute principles.

We will get along even better than I thought. Your response implies that in some situations you would be willing to accept any price in order to achieve your end, in others not. Depending on the situation. Yes. That is an important thing to know about one's self."

The stranger unfastened his gaze from me and stared out into the Void. He was apparently occupied by something in particular, a particular one of the cosmoses, misshapen and throbbing as if it were about to explode. He looked at it with fascination. Then he turned sideways. He was so thin that he practically vanished, appearing as only a black line. "Have you wondered," he said, "whether it is possible to imagine everything that will ever exist, or whether some things lie beyond our ability to imagine them?" I nodded. "And the set of all possibilities being infinite, as it is," he continued, "if there is even a fraction of possibilities we cannot imagine, then there is an infinite number of possibilities we cannot imagine. So, even with infinite power, we might be surprised by what transpires in the future. Would you agree?" The tall stranger turned towards me again, cocked himself at an angle, and looked at me with an odd expression.

"Yes."

"These universes you've created," he said, and gestured at the quivering spheres and ellipsoids flying about. "Many of them will end in tragedy. Or I should say, the animate matter you fill them with, the intelligent beings, will twist and suffer and meet unhappy ends." He smiled.

"I have no intention of that," I said. "I would not allow that to happen."

"I am sorry if what I've said disturbs you."

"I command you into nonexistence," I said.

"I'm afraid you cannot do that." As tall as he was, the stranger grew taller, as if he had been crouching. "The glittering multitudes," he said. "So many little lives, amounting to nothing. I ask you: What is infinity multiplied by zero? It is hardly worth our discussion . . . Give my regards to your uncle and aunt." The stranger bowed. Then he and his beast, looking back at me with its incessant grin, moved off through the Void.

Second Thoughts

In my anger, I smashed thousands of nascent universes. Some, I strangled the space out of them, leaving dry husks of nothingness. From Void to Void. Others, I spewed in so much energy that they exploded in a soundless catastrophe. Some universes I hurled at other universes, splattered them into each other. I ripped apart space. I scattered geometry. I crushed and destroyed. Never before had I felt such emotion, and the Void seethed with my anger, the Void's music devolved to a screech of clashed chords.

What are you doing? cried my uncle. He stooped to pick up pieces of the fractured universes. You have frightened me, and your aunt as well. The two of them rushed about as if looking for somewhere to hide, then cowered a distance away, each trying to shelter the other.

I would, of course, never do anything to harm Uncle Deva and Aunt Penelope, but I found myself behaving without any thought. I was pure action, and

Mr g

I watched myself wreak havoc as if it were another being moving about and crushing alien creations. I was outside myself. How long this went on was difficult to gauge. Eventually, my fury softened. Looking about, I could see that I had annihilated many of the universes I had made. But many more remained, growing larger. I had not destroyed everything.

I told Uncle and Aunt about the stranger. The arrogance, said Aunt Penelope. He had no right to come here, and certainly not in that manner. Just let him show up again. You should not be discouraged.

I don't know, I said to my aunt. Perhaps you were right. I should have left everything just as it was, in an infinite nothingness. I do not want my creations to end in tragedies. I should have left things as they were.

Tragedies? said my aunt. Are you referring to the creation of animate beings in your universes? Listen to me, Nephew. First, you have not made animate beings. So far, you have made only empty cosmoses. And secondly, even if you do create animate beings, you do not know that they will suffer tragedies just because that swaggering desperado said so. You forget your power, Nephew. You made those cosmoses. If you choose to, you will make animate beings. And you will make them as you wish. Have faith in your creations. Yes, yes, said Uncle Deva. Have faith. Your aunt and I stand behind you. Don't we, Penelope? Absolutely.

I looked out into the Void, at the billions of cosmoses whizzing about, and I imagined populating each of them with matter, both animate and inanimate. I imagined atoms and molecules. I imagined gases and liquids and solids. I imagined silica and soil, atmospheres, chemical elements, oceans and lakes, mountains, forests, great lumbering clouds, electrical impulses in space, movements of ions, gelatinous membranes, bacteria. I imagined brains, some made of matter and some made of energy. I imagined intelligent creatures. And *their* creations. Their cities. I tried to picture the future. Would my living creations suffer and writhe in some agony? Was it necessarily so? Or would they have only pleasure and joy? I felt the future, but I could not hear it. I listened. Could I hear the voices of the trillions of creatures who might come to be? Could they tell me of life? Could they tell me of suffering? But I could not hear them. All I could hear was the soft adagio of the Void. I felt the future, but the future did not exist. I gazed at the billions of universes, fraught with their emptiness and possibilities, and I wondered. Perhaps I should make only nonliving matter. That would be simpler, and safe. But could I limit my productions to inanimate matter? I could make whatever I wished, but could I be certain about the subsequent movements of each atom once made? Could I be certain that trillions of dull and dead atoms could never combine and give rise to a thing that had life? And there were so many worlds.

Some Organizational Principles

May I give you some advice, Nephew? said Aunt Penelope. The three of us had been wandering about in the Void for some time, talking about how our existence had changed and sweeping up bits of debris still lying about. Don't give Him advice, said Uncle. He doesn't need our advice. Hush, said Aunt Penelope. I am entitled to give advice to my nephew. If it is not to your liking, then give Him your own advice. I would be careful, said Uncle. Do you really— Aunt Penelope cut off Uncle with one of her looks. But now that she'd been regularly combing her hair, she did not appear nearly so fierce as she once had. Still.

Aunt Penelope took me aside, leaving Uncle by himself. Now, I want you to listen to me, she said. This is no criticism. Your uncle and I have always been impressed with you. But we are your elders, and we do notice what goes on around here . . . You shouldn't do things with such haste. You rush into things. Slow down. Take your time with this project.

I hadn't been aware I was rushing, I said to my aunt.

All these things flying about? said Aunt. You made them so quickly. Why don't you concentrate on just *one* of your universes and see if you can do a good job with it.

That's excellent advice, said Uncle Deva, standing some distance away.

Which one would you like? I asked my aunt. It wasn't really a serious question. There were quadrillions of spheres and hyperboloids flying about, by now having inflated to at least 10^{17} times larger than they were just a few moments ago. This one, said my aunt, and she suddenly reached up and caught one of the spheres flying past. Work on this one. We have confidence in you, your uncle and I, and we are certain that you can do well with it. Now that you've started this project. Just take your time, that's all I'm suggesting.

Perhaps my aunt had given me decent advice. The universe in question was nearly spherical in shape, spinning slightly, and it was inflating with a rabid determination. The first thing I did was to slow its expansion. There, said Aunt Penelope, at least now we can examine it. We? said Deva. Let Him examine it on His own.

Mr g

I should mark this one, I said, so that it will not get lost among the others. I pinched the universe very slightly, making a small dip in its middle. Interrupted in its flight and caught, the thing sat there quietly.

We must give it a name, said Uncle Deva. Everything has a name. Something with a lilt. Something pretty. Why not call it Amrita. Or Anki. Or Aalam.

Oh mush, said Aunt Penelope. You're being sentimental. And you can't name an entire universe anyway.

Of course you can, said Uncle. A name expresses its essence. A name gives a thing character, personality.

But a universe doesn't have a personality, said Aunt. As I understand it, a universe is a . . . well, a totality. A universe is everything that is, as far as the inside of the thing.

But we're on the outside, said Uncle Deva.

If we have to name it, said my aunt, at least call it a number, not one of those mushy things you said.

A number! cried Uncle. That's so impersonal. Numbers are so remote. What do you say, Nephew?

I looked at the pinched cosmos, still held firmly by my aunt as if she were afraid it might go whizzing off any moment. It seemed pretty featureless to me. But perhaps it would grow into its name. All right, I said. I'll call it Aalam-104729. So be it.

104729? said Uncle. What a random number.

It's the ten thousandth prime number in base ten, I said. I won't forget it.

You see why I wanted a name? said Uncle. Now put some spirit into the thing.

Look who's telling Him what to do, said Aunt Penelope. A moment ago, you didn't—

Everything must have a spirit, Uncle said to me. Do it however you want, just give it a spirit. And use *feeling*. You've made something grand, but it will be grander if it has feeling and beauty and harmony and—

Deva, I've never heard you talk so much, said Aunt Penelope. This discussion is wearing me out. I'm going to sit. Where's my chair? Where's my chair? Uncle Deva shuffled off and fetched the chair, which he had named Guptachandraha. My aunt, clutching Aalam-104729, went over to her chair and sat down. She stretched out and sighed and began mumbling: First it's this, then it's that. If it's not one thing, it's another, not one thing, it's another, not one thing, another. Her mumbling gradually tapered off, and she pretended to fall asleep.

I have to think about this, I said. I'm afraid if I put in spirit and feeling, before anything else, the thing is going to get all jumbled up and confused and end up in chaos. It needs to start off with some *organizational principles*.

OK. OK, said Uncle. It's your project. Organizational principles. OK. We will leave you to it. Do tell us when you are finished with the . . . organizational principles. Leave Him to it, he said to Aunt Penelope, who was still pretending to sleep. Uncle walked over and extricated Aalam-104729 from her grip and gave it to me. Organizational principles, Uncle said once more. Take your time with it, said Aunt Penelope. That's all I ask.

I generally try to be everywhere at once, but I moved to a place in the Void where I could be alone. I meditated, and I entered the pinched universe and looked about. It was empty of course. I imagined moving in various directions in space, and I also imagined traveling forwards and backwards in time, and I decided that I wanted my universe to be completely symmetrical in time and in space, so that one place and one moment should be the same as any other place and moment. This was by far the simplest cosmos I could make, and I wanted my first universe to be simple. Symmetry of position and moment. This was my first law. And I remade Aalam-104729 to obey this first law. For a few moments, the universe quivered and murmured, and then it was still. The first law seemed good to me.

But then I began considering future and past. Inside Aalam-104729, I wanted to know clearly that the future was different from the past, so that any intelligent being could tell that *things were happening*. Wasn't that precisely the point of waking from my slumber, to make things happen?

So I remade the energy in my universe so that it was all concentrated in a near-perfect order, a razor-sharp contour of energy. Almost at once, the razor of energy began fraying at the edges, loosening, dulling, and diffusing away, and this was good because now there was a definite future and past. At any moment, the past was the direction of time with greater sharpness and shape, and the future was the direction with less. I was pleased.

Then I made a second law. There would be no absolutes in my universe, only relatives. In particular, there would be no such thing as absolute stillness in Aalam-104729. I wanted the only point of absolute stillness to be Myself. If something appeared still from one perspective, from another perspective it would be in motion. If a material object changed its motion, going from one motion to another, everything should remain the same, with no reference point of stillness to say that one motion was any different than another. This second law was a principle of symmetry, like the first, and there was an artistic beauty in it, and it was good. Or—if a principle could not be deemed good or bad—at least it was satisfying, it seemed in harmony with the music of the Void.

The second law necessarily tied time and space together, since motion involved the two. A particular period of time would signify a particular distance in space, with the proportionality between the two being a fundamental speed of the universe. This relationship between time and space was also beautiful and good.

Was I acting too hastily? I wondered if Aunt Penelope was watching. Even though I was inside Aalam-104729, I could look outside, because I could look everywhere, and I could see Aunt and Uncle far off in the Void, paying no attention to me. Uncle Deva had somehow installed himself in my aunt's chair, stretched out as if he meant to spend quite a long time there. Meanwhile, she was swatting at him, shoving and pushing in an attempt to dislodge him.

With Uncle and Aunt thus occupied, I made a third law: Every event should be necessarily caused by a previous event. I did not want things happening willy-nilly in my new universe. Events without cause would lead to a reckless cosmos, a universe ruled by chance. According to my third law, for every event, there would be a previous event without which it would not have happened. And that previous event would also require and be determined by a previous event, and so on, back through an immense chain of events to the very *first event*, which was my original creation of the universe. This law was also good. It prevented pandemonium. It bestowed Aalam-104729 with causality. It bestowed logic and rationality. And it connected everything. Cause-and-effect relationships would spread out from every event to every other event, even to multiple subsequent events, ripple through the cosmos, and bind the totality of being in a web of interdependence and connectedness. Even the smallest event would be linked to other events. Wasn't this a kind of spirituality? See, I wanted to tell Uncle Deva (who was at that moment still scuffling with Aunt Penelope over the single chair in existence). Rationality and logic can be spiritual.

What's more, there was still plenty of room for the mysterious. Because even if a very intelligent creature within this universe could trace each event to a previous event, and trace that event to a previous event, and so on, back and back, the creature could not penetrate earlier than the First Event. The creature could never know where that First Event came from because it came from outside the universe, just as the creature could never experience the Void. The origin of the First Event would always remain unknowable, and the creature would be left wondering, and that wondering would leave a mystery. So my universe would have logic and rationality and organizational principles, but it would also have spirituality and mystery.

Mr g

Three laws. I floated about the interior of Aalam-104729, squeezing the vacuum here and there to see if the laws held, and they did. No loose parts or inconsistencies. I was satisfied with what I had done. More than satisfied. In retrospect, creating a principled universe did not seem so difficult. I had been concerned for no reason. I was eager to make a fourth law. Perhaps I'd do a dozen. Or two dozen.

What should I do for my fourth? Uncle Deva wanted harmony. My symmetry principles were already harmonious, but I could do better. I divided the ubiquitous energy into parts, each with its corresponding force, and I ordered the forces in a progression from the weakest to the strongest. All right. Harmony. I decreed that each force in the progression was stronger than the preceding force by a constant ratio, like an even-tempered musical scale. Done. What could be more harmonious! But, almost immediately, the universe began writhing and straining. Space fissured. Pieces of emptiness screamed through the tears. Shortly thereafter, the universe turned inside out and was gone, and I found myself standing beyond in the Void. Evidently, the fourth law was not compatible with the first three. The constant ratio of forces, although beautiful, contradicted the even greater beauty of embedded relativity. Looking about, I saw that, fortunately, Aunt and Uncle were nowhere within sight. Quickly, I caught another universe of roughly the same size and shape as Aalam-104729, pinched it slightly at its midsection as I'd done before, and gave it my first three laws. Three it would be, and no more. I wouldn't make that particular blunder again.

A Soul for the Universe

When Aunt Penelope and Uncle Deva saw the cosmos I'd made, with its three laws, they were not displeased.

Well, what we got? said Uncle, looking more rumpled than usual. He held the universe up and squinted at it from all sides. Although it appeared nearly the same as before, it had a new heft, he announced, and it vibrated with a higher frequency. Yes, said Uncle, the three laws seem to be agreeing with the thing.

It's because He's taking His time, said Aunt, like I told Him to do. You take your time, and you can do good work. You rush into things, and you might destroy a whole universe. What a pity that would be.

Uncle Deva passed the new Aalam-104729 over to Aunt Penelope, who began her own inspection. She rolled it over on its side, turned it upside down, spun it around. It was still expanding, getting bigger every moment. She nodded her approval. So, Nephew, she said. What's next?

It's still empty, I said. Perhaps it's time to start putting things into it.

If I may make one last suggestion, said Uncle Deva. You say that your universe has a spirit. I don't follow all of that folderol of causal connections and so forth. You always defeat me in those kinds of explanations. Be that as it may, I would be grateful if you give your universe a *soul*. You need to make sure that everything in the universe is connected not just to other things, but to *you*. You are the Maker, after all.

I don't feel that's necessary, I said. I know I'm the Maker. But there's no reason my creations need to know it. You know it. Aunt Penelope knows it. That's sufficient.

Don't be modest, said Aunt Penelope. For once, I agree with your uncle. You are the Maker of everything. Your creations should understand that. They should have some awareness of you and your infinities. And it's not just about you. It's about our family, all of us here in the Void, our reputation. You're an artist, Nephew. Deva and I appreciate your artistic work, but that's a small audience.

Aunt Penelope, please. I haven't decided whether I'm going to make any living creatures period, much less *aware* creatures, much less creatures aware of

Me. It might be a comfort to be unaware. I might decide to make only inanimate matter.

What a waste! said Uncle. To make such a beautiful universe filled only with inanimate matter? It would be *boring*. Boring, I tell you. Am I the only one who thinks it would be boring?

It would be boring, said Aunt Penelope.

Yes, I said. It might be boring.

Then we are agreed, said Uncle. There will be animate matter with intelligence, and there will be an immortal soul in each living being, connecting it to you.

Wait a moment, I said. Only we, and the Void, can be immortal. Immortality does not exist in Aalam-104729. The thing has a direction of time, caused by the dulling of its energy, and everything in it will eventually dissipate. Nothing lasts forever in Aalam-104729, or in any of the universes I have created. I will consider a soul, but it cannot be immortal. It must follow the direction of time, like everything else. It must gradually decay and disintegrate. We cannot begin making exceptions to the rules here and there, helter-skelter, or we'll end up with chaos again. Let me consider this . . . Maybe in the life of each creature I will allow a brief recognition of something vast, a flash of Me, a hint of the unchanging and infinite Void.

And then those creatures will pass away? said

Deva. Dissipate and die? And their souls with them? At least let the souls come back in new bodies. Otherwise, it is so sad.

There you are getting mushy again, said Aunt Penelope. What do you know of sadness? What do any of us know? Sadness may not even exist. Let's take a walk. I feel like stretching a bit.

Matter

All of us—Uncle Deva, Aunt Penelope, and I—were feeling protective of the fledgling Aalam-104729. We didn't want to leave it alone among the zillions of other universes flying about, so we carried it with us on our stroll through the Void. Although its mass was infinite, the infinity was a small infinity, so it felt like nothing at all.

My aunt moved ahead, poking at folds of the emptiness as was her habit and stopping to collect little scraps of Void for some private use later on. She was in fine form. We could hear her commenting and exclaiming to herself. Uncle and I lingered behind. He had always been slower than she, one of their many marital incongruities. After every long sleep, she would leap up with plans for a new sightseeing excursion through the Void, while he would roll over, look groggily about for a few moments, and go back to sleep. Ironically, my aunt was the more patient one.

She would take any amount of time for even small things (except her own appearance), whereas Uncle worked in broad strokes and easily grew upset with details. He was the idealist, she was the more practical member of the pair.

You've got to put in a soul, whispered Deva. You heard what your aunt said.

I'm thinking, I'm thinking, I answered. Dear Uncle, can we not talk about the soul for a while and just enjoy our walk? Listen to the music. At that moment, a playful scherzo was resonating throughout the Void. Listen. Yes, of course, said Uncle Deva. For a while. I have no idea where your aunt has got to.

Unmeasured time passed.

I am going to make matter, I announced. Inanimate matter. And then? asked Uncle. You are not going to stop with that? There is plenty of time to decide what to do after that, I said. Do you mind if I attend to this now? Uncle shrugged. I'll be back in no time, I said. In fact, you won't know I'm gone.

I entered the new universe again and took stock. Matter. At the moment, Aalam-104729 contained only pure energy. But my two symmetry laws already guaranteed that matter could be created from energy—in fact, required it—so all I needed to do was to specify the parameters of a few basic particles. This one spins this much, that one spins that much, this one

46

responds to this force, that one to that force, and so on and so forth. Done.

Immediately, matter appeared! In fact, matter exploded. Matter burst into being with a vengeance, as if it had been languishing in a frustrated state of potentiality for eons of time and was finally given the opportunity to exist. Electrons and muons and taus, top quarks and bottom quarks, squarks, gravitons, photons, neutrinos and neutralinos, gluons, W and Z bosons, axions, photinos, winos, and zinos. And with matter, of course, came antimatter: positrons, antimuons, antiquarks, et cetera and anti et cetera.

At every point of space, the hillocks and basins of energy gushed forth with matter. Some of this matter instantly annihilated with antimatter to create energy again, which in turn spit forth new matter, so that there was a continual give and take between the two. Energy begat matter which begat energy which begat matter. It was a spectacle.

The photons in particular sometimes took the form of an oscillating wave of electrical and magnetic energy. I decided to call such a thing "light." Where photons flew about in abundance and collided with other matter, there was light. Where photons were absent, there was darkness. Thus, when I created matter and energy, I also created darkness and light, and I decided that these things were also good, although I was not sure at the moment exactly what they would be good for.

So now there was matter and energy. And as I was watching, the universe grew and it cooled. The average energy of each particle diminished, and eventually some of the elementary particles began to coalesce with one another to make larger particles and masses. I could imagine, in the future, the formation of atoms and molecules, ripples of electromagnetic energy streaming through space, vast clouds of gas condensing under the gravitational force, spiral-shaped galaxies studded with bright balls of gas. Inside these spheres, roiling with nuclear reactions, new elements would be formed—carbon and oxygen, sulfur and magnesium. Great diffusions of neutrinos and light. And then titanic explosions, spewing more elements into space. And I could imagine vast disks of gas rotating around embryonic stars, elliptical orbits of comets, condensations of matter into rocky planets of silica and iron, or gaseous planets of hydrogen and helium, icy planets of frozen methane, molten planets of liquid sulfur, planets in retrograde motion, seething magnetic fields accelerating matter to maximal speeds, atmospheres of gaseous sulfur dioxide, oceans and mountains and silicon lagoons. In time, it would all come to pass. And all of it dumb, inert matter.

The Stranger Returns

Aunt Penelope emerged from somewhere, beaming with a healthy exhaustion. She was practically bent over, shouldering the piles of emptiness she'd gathered. Nothing like it, she said. Nothing like it. She looked at Uncle Deva and frowned. If you could just move a little faster, she said. Look at you. You are so confoundedly slow. And *you*, answered Uncle, tear through the place like . . . I don't know what you're like, but you do it. Can't you stop and listen to the music? And can't we walk *together* for once? He kissed her. Deva! she said. Not out here in the open. She gave Uncle some of the odd-shaped patches of Void to carry for her, sighed, and began walking beside us.

I've made matter, I said casually.

Really! said Aunt Penelope. She picked up Aalam-104729 and shook it. It rattled. Yes, she said, there are pieces of stuff inside it now. Congratulations again. All of these congratulations are getting tiresome, I should think. Perhaps we've had enough congratulations for a while.

It's time to put in a soul, said Deva.

But it's only inanimate matter, I said. I haven't—

We were interrupted by a howl, followed by a snicker. Then Baphomet appeared in front of us, grinning as usual. The beast jumped up and down and cartwheeled, keeping its stupid gaze fixed on me.

"What do you want?" said my aunt. "Off with you."

"Him," said the creature, pointing at me. "My master wishes to speak to Him. My master is waiting over there." The squat beast gestured in a particular direction of the Void.

"You are an outrage," said my aunt. "My nephew does not answer to summons. My nephew is the summoner."

The beast turned to stare at Aunt Penelope with its hideous grin, snarled, and began laughing. "You forget yourself, Penelope." Then the creature did a somersault in such an extraordinary manner that its grin seemed not to move at all while the rest of it tumbled through the Void and returned to an upright position. "My master does not like to be kept waiting," said the beast. "Suit yourself." It snarled again and went trotting off.

"This is intolerable," my aunt said to me. "This creature—whatever it is—speaks to me like . . . Oh, it is intolerable. Disgraceful."

"I observe that the making of laws has begun," came a voice from behind us. And there was the stranger, twice the height he had been before. He bowed to Aunt Penelope and Uncle Deva. "May I see it," he said, and somehow gained possession of Aalam-104729 without moving. "This particular universe you have chosen among all the others. I wonder why. But it makes no difference. This one it will be." He studied it closely and sniffed it. It seemed to shudder in his grasp.

"That does not belong to you," said Uncle Deva.

"It belongs to all of us," said the stranger. "It is . . . how shall I put it . . . It will provide the path whereby we complete ourselves, make ourselves more than what we are now."

"Destroy this abomination," Aunt Penelope said to me. She turned to the stranger. "Who are you?"

"I am called Belhor," said the stranger. "You may also call me Fedir or Belial or another name if you wish. And I apologize for any offense given." The stranger bowed again and offered the pinched universe back to Aunt Penelope. "As I told your nephew earlier, you have very pleasant accommodations in these regions. And I must further compliment you on the lovely music."

"You are not entitled to exist," said my aunt. "My nephew did not make you."

"Oh, but He did," said Belhor.

"Oh, but He did, He did, He most certainly did," said Baphomet, and the beast erupted in laughter and performed two somersaults and a bow. "That is the most delicious part of it all."

"What is your business here?" I said. "Although I have unlimited power, I have limited patience."

"Well spoken," said Belhor. "I would like to discuss your laws with you, the ones you are devising for your new universe."

"There is nothing to discuss," I said. "The laws have been made."

"Ah," said Belhor. "But the laws you have made are only the physical laws, governing elementary particles and forces. Am I not correct? At some point in the future, intelligent beings will exist in your universe, and it is the laws governing the behavior of such beings I would like to discuss. I believe that we all share the view that animate matter is far more interesting than inanimate matter."

"Animate matter will be governed by the same laws as inanimate matter," I said.

The stranger laughed. "For someone as powerful as you . . . I am afraid it is not so simple. As we said earlier, the mind is its own place. None of us, and especially you, should underestimate the complexity and subtlety of a mind, once formed. Do you mean to say that every action and every thought of an intelligent creature in your universe will be determined by preexisting physical laws?" I nodded. "In that case," continued Belhor, "your intelligent beings will have no independence of movement or thought. They will be completely controlled by you, or rather, by the physical laws that you have created, which is tantamount to the same. In fact, could we not say that their lives are already prescribed? Am I understanding you correctly?"

"Oh, what fun this is," said Baphomet. "I'm so glad I came." The beast turned to Aunt Penelope, grinning its unceasing grin. "Didn't I tell you that my master was not to be trifled with? Didn't I? Didn't I?"

"I can see that you have been thinking about this," I said to Belhor. "Yes, what you say follows logically. The thoughts and actions of animate matter, if such matter comes to exist, are already determined by the laws I have made and by cause-and-effect relationships. That is my intention. If there are rules, there are rules, and no exceptions for animate matter."

"So you wish for total control over your creations, animate as well as inanimate," said Belhor.

"Nephew, you must do away with this monster," Aunt Penelope said to me. "He is bullying you."

"No, Aunt. This discussion interests me."

Belhor bowed. "Thank you for your respect," he said. He looked at me for a few moments without speaking. "And I say to you, with equal respect, I believe that you are allowing your ego to get in your way. Why do you require such complete control? Do you not trust your intelligent creations to act on their own, without your supervision? Do you think they may do something to embarrass you, or something you consider unseemly or unworthy?"

"It is not a matter of trust," I said to him. "It is a matter of self-consistency." I found myself concerned, and I wanted to examine my concern, to turn it around in my mind and to explain it with clarity. "Surely you can understand that I do not want the universe to be a hodgepodge of contradictory rules and events. Where would that lead? All of us appreciate a certain amount of ambiguity and subtlety, but there must be limits . . . Animate matter should be subject to the same laws and rules as inanimate matter. Since those laws completely determine the behavior of inanimate matter, they also completely determine the behavior of animate matter."

"Then animate matter is no longer interesting," said Belhor. "I am disappointed in what I have heard. And, for the moment, I have no further business with you."

"Oh, my master is disappointed," said Baphomet, who had snatched the universe from Aunt Penelope and was tossing it up and down like a plaything. "My master is sooooo disappointed, and it is a very bad thing to disappoint Master Belhor."

"Please excuse Baphomet," said the stranger. "He lacks manners, and he gets excited. Now, we will be off."

The Universe
Nurtures Itself

What a disagreeable fellow, said Uncle Deva. And arrogant. Did you see how he strutted about, as if he owned the place.

He has a bad odor, said my aunt. I'd like to know his origins. If I had the power, I would—

I fail to understand why we have to tolerate that disagreeable fellow, said Uncle. He is causing discord and bad feelings, after eons and eons in which everyone here in the Void got along beautifully.

Quite right, said Aunt. Then she began looking about. She frowned. Where is our universe? Did that abominable creature, that Baphomet, take it with him? We've lost the universe. The last time I saw it, the beast was batting it back and forth, acting like he was going to eat it.

My uncle and aunt began pacing about in search of Aalam-104729, she moving much more rapidly than he, of course. They peered under layers of nothingness, thrashed at diaphanous tendrils and veils of the emptiness, listened intently as if they might hear the slight rumblings and regurgitations of the infant universe.

Where is it? muttered Aunt Penelope. I will strangle that monster myself. Both of them. Where could the universe be? It was not so large, really. Still a small universe. Not so big. Not so big. Where is it? Where is it? Aunt Penelope was walking around in great circles, periodically returning to where Uncle stood, glancing at him with a frown, then setting off again. Meanwhile, my uncle toddled about in no particular direction, confused and concerned.

I am not sure how much time passed in this state of affairs, as I was brooding over the conversation with Belhor. Finally, my aunt spoke up: Nephew, don't just stand there. Might you take it upon yourself to provide some assistance?

I, who could see everything at once, perceived that Aalam-104729 was some distance away behind a fleeting hillock of nothingness, lying on its side as if it had been carelessly tossed away by the ever-grinning Baphomet. There it is, I said.

Oh, cried Uncle, and he shuffled over to the spot and scooped up the castaway universe and cradled it against himself. I am getting attached to this little universe, he whispered. I am embarrassed to say that I am becoming attached to it. He held the universe for a while and then placed it gingerly in a fold of the Void.

Attachment leads to disappointment, I said. I liked to ruffle Aunt and Uncle now and then when I could.

Yes, yes, I know, said my uncle.

I found myself still agitated, as I was after Belhor's first intrusion. But when I gazed upon Aalam-104729, a rosy plump ball, already considerably larger than when I had last noticed it, I felt a sense of calm and hopeful expectation. So much was possible with a new universe. And I now understood Belhor's comment that this universe would make all of us more than we were now. For vast epochs of unmeasured time, we had slumbered, we had existed in beautiful but vacuous nothingness. In retrospect, we had been colossally *bored*. This plump, expanding sphere, ripening with possibilities, could change everything. It was smaller than us, but also bigger. And there it sat quietly in a small furrow of the Void, seemingly none the worse for its rough handling by Baphomet.

It will want some tending, said Aunt Penelope.

I don't think so, I said. I've given it several laws and a few quantum parameters. Cause and effect. I think it can fend for itself.

Please, said Aunt P. The thing is so . . . Tender, said Uncle Deva.

Mr g

To oblige my uncle and aunt, I entered the universe again and looked about. Indeed, the cosmos was humming along, in no need of my help. Since my last visit, the universe had cooled, and more kinds of particles were able to sustain unions with one another under their mutual attraction. I was intrigued by the varieties of things and effects. Triplets of quarks had combined to form neutrons and protons. These flew about at a ferocious speed, surrounded always by an ultraviolet haze of soft gluons and occasionally emitting gamma rays as they ricocheted off other frenetic nuggets of matter. Particles spun about their internal axes. Particles swerved in magnetic fields. Particles careened and accelerated and annihilated into pure energy. Here and there, bunched pockets of electrons or positrons would form, slight deviations from the mean density, and these unbalanced charged regions oscillated and vibrated in response to the electrical attractions and repulsions between them. Following my laws for the electromagnetic force, each such quivering of charged particles unleashed a flood of polarized photons with kaleidoscopic colors, creating a display far more spectacular than the evanescent veils of the Void. There were cascades and blooms of light, spiraling helices of energy, resonant oscillations of quark clouds. And the most eerie sounds: ultra-high-frequency moans and rips and dissonant crescendos as the gaseous plasma filling up space shuddered with each passing shock wave and compression of energy. Necessarily, there were small valleys and summits of matter and energy, inhomogeneities. The force of gravity struggled to strengthen these scattered accumulations, but the particles were so energetic and hot that gravity seemed almost nonexistent. That situation would eventually change as the universe expanded and cooled further. Between condensations of matter, the vacuum was constantly erupting with pairs of particles and their antiparticles so that there was not a single dollop of space that could be said to be sleeping. Indeed, the "vacuum" of space seethed with the creation and annihilation of new particles. In these early moments of the new universe, every pinprick of mass flew about with nearly the same speed as the photons, the maximum speed allowed by the laws. Space was a buzzing blur of subatomic particles, whizzing about at fantastic speed in crisscrossing patterns, zipping about and deflecting and colliding with one another. Energy fields lay across the cosmos in vast, floppy blankets, slightly shuddering as they created each new particle or absorbed other particles into their folds. And in every volume of existence, quantum physics held sway. Particles acted as waves, waves as particles. Alternate realities shimmered at every position of space. Matter and energy appeared and disappeared, merged into each other, and exchanged identities. And at the tiniest sizes of somethingness, quantum fluctuations and gravity conspired to tessellate the very geometry of existence.

It was exhilarating. It was glorious. It was more than I had imagined. At the same time, it was all entirely logical. All of it followed inexorably and irrefutably from the few laws I had laid down. I had to do nothing but sit back and watch as the cosmos unfolded in time.

The Quantification
of Reality

Time. As yet, time was unmeasured and unmeasurable. But that was soon to change, with the formation of hydrogen atoms.

As Aalam-104729 continued to expand and to cool, there came a point at which it was sufficiently tepid that electrons could be captured and held by protons to form atoms of hydrogen, the simplest of atoms. In each hydrogen atom, a single electron orbited a single proton. Hydrogen atoms were my first atoms. They were lovely. Some were spherical, others ovaloid or dipoloid, depending on the quantum state of the orbiting electron. Patterns within patterns within patterns, all perfect as the number π and precisely determined by the few quantum rules I had given. The atoms glowed as their revolving electrons emitted photons. They faintly hummed. And the atoms gave matter a sponginess, a kind of cushiony texture it did not have before.

Mr g

Most importantly, hydrogen atoms served as the first clocks. The light emitted by these atoms vibrated with a precise regularity in time, always exactly the same, each vibration being one tick of the clock. Peak and trough and peak and trough and peak and trough—tick, tick, tick, tick, tick, tick. Now any duration of time could be measured by how many ticks of an atom of hydrogen. In these terms, Aalam-104729 was at this instant 4.52948×10^{29} atomic ticks old. The first neutrons and protons had begun forming at about 2.5×10^9 atomic ticks after the birth of the universe, the first atoms at about 3×10^{28}. I was somewhat surprised to realize that a great deal of time had already elapsed, at least in terms of the beats of the atoms of hydrogen.

Now we had clocks. Now time not only existed, but it could also be quantified, it could be measured, it could be carved up into pieces equal to the quantum throbbings of atoms. Now we could do far more than say that something happened in the past. We could say precisely how far in the past. And the duration of happenings and events, the time elapsed between A and B, could be assigned a definite number. The concepts of fast or slow, lazy or brisk gained a definite meaning. At last, I could measure the interval between Aunt Penelope's great heaves and snores as she lay sleeping (typically 10^{20} atomic ticks when she retired in a good mood, less when she was disgruntled). One of Aunt P's interminable speeches, advising me to do this or that, often took 10^{21} or 10^{22} ticks. And one of Uncle D's leisurely walks around the Void occupied between 10^{25} and 10^{26} ticks. (Compared to these events, my thoughts were so rapid as to be almost instantaneous, occupying a mere one-millionth of one-trillionth of one-trillionth of a single atomic tick.) I hesitated to calculate exactly how long I'd been doing absolutely nothing, how long all of us had slept in our torporous amnesia.

As time and space were connected by the speed of light, the quantification of time naturally led to a quantification of space. Now any length could be measured in terms of the distance traveled by a photon of light during one tick of an atomic clock. In these terms, the diameter of a neutron was one-hundredth of one-millionth of a light-atomic-tick. The diameter of an atom was a hundred thousand times larger. The diameter of the entire universe, judging by how long it took a photon to traverse the distance, was 9×10^{29} light-atomic-ticks, and growing larger each moment.

Delighted to have a reliable method to quantify reality, I immediately set about measuring everything I could find. I measured the size of certain quark condensations: 10^{-7} light-atomic-ticks. The average size of a matter inhomogeneity: 10^{27} light-atomic-ticks. The time for a particular basin of antimatter to annihilate with matter: 1,003 atomic ticks. The time for the universe to double its size: 10^{30} atomic ticks.

Uncle Deva was appalled that I might now lay ruler and clock to the Void. He appreciated what I had achieved, he said, but I was going too far. Too far? Tell me, said my uncle, what do you know about a thing when you know precisely its size and its duration? You know precisely nothing, that's what you know. But how can you compare the thing to other things? I protested. Why should you compare? said Uncle. Each thing possesses its own special essence, which has nothing to do with anything else. Understand the essence of a thing, said Uncle, and you know everything you need to know. And I guarantee that the essence is not how many what-you-may-call-it atom flicks you've got. No sir. You're only fooling yourself.

Aunt P looked suspiciously at the loose hydrogen atoms I'd brought back to the Void. Don't you dare measure me with those gadgets, she said. But I was only . . . No ifs, ands, or buts, said my aunt. I am *unmeasurable*, and I aim to stay that way. Period. I don't want some half-witted creature in some universe or other quoting my measurements. Just don't bring those gadgets into the Void. Let them stay where they are. Amen, said Uncle.

It seemed that I was the only one who took pleasure in the new clocks and rulers.

Galaxies and Stars

Bound by causal necessities, requiring not a single touchup or tinker from me, events in Aalam-104729 proceeded on their own with an impressive inevitability. As the universe continued to expand, its material contents cooled further and further. The brilliant displays of light slowly dimmed. And the attractive force of gravity began to dominate and reshape the terrain. Whereas before, small condensations of matter would quickly evaporate under the high heat, now they grew larger and denser. Lumps of material, most of it hydrogen gas, began to condense here and there. In the past history of the universe, matter had been rather evenly spread about, but now there were ridges and valleys, arches, amorphous aggregations, all bunching themselves up into ever denser bulges as each particle of mass gravitationally attracted other particles. The smooth, almost fluid topography of matter before had been beautiful, but these architectural constructions were even more beautiful. There were linear filaments. There were sheets. There were hollowed-out spherical cavities. There were ellipsoids and spheroids and topological hyperboloids. Great clouds of hydrogen gas swirled and flattened and spun out spiral wisps and trails. And within these spinning galaxies of matter, smaller knots of gas formed, collapsed on themselves, and grew denser and hotter—in opposite fashion to the rest of the universe, which was thinning and cooling.

After 10^{31} ticks of the atom clocks, a wondrous new phenomenon occurred. Each knot of gas in each galaxy had reshaped itself into an almost perfect sphere, which grew hotter and hotter as it contracted under its own gravity. Eventually, the heat was so high in these globules of mass that their hydrogen atoms began to fuse with each other to make helium atoms, the next simplest element after hydrogen. In every atomic tick, trillions of these fusions occurred, releasing vast quantities of nuclear energy. Just moments before, these ubiquitous spheroids of gas had been only dark pebbles in the darkness of space. Now they were shining, bursting forth with energy. The first stars had been born.

I remember where and when I beheld the first star in the universe. Actually, I was taking a long walk in the Void with my uncle, listening to him hum his favorite piece of music, a dissonant screech of a tune, when I noticed something change inside Aalam-104729. A tiny light shone in one of the billions of dim galaxies. (Uncle has since named this particular galaxy that spawned the first star Ma'or, and the first star itself Al-Maisan.) Looking more closely, I saw that a single globule of mass, less than one-trillionth the size of the galaxy, was producing all the light. Such a tiny speck in comparison to the galaxy. Yet it was unmistakable. In the black reaches of space, this single pinprick of light could be seen. It gleamed and it pierced. The ultraviolet radiation emanating from this dollop of matter raced outward in all directions into the surrounding gas and disintegrated the nearby atoms, tearing the electrons from their protons. As the electrons reunited with the protons, cascading down to lower and lower energy levels and emitting light as they did so, a spherical cocoon of gas around the star began to glow with violets and yellows and oranges and reds. The star now appeared as a point of ultraviolet light at the center of a soft glowing cloud of many colors.

Then, one by one, other stars ignited. There. And there, and there. Now there were hundreds, thousands, millions of ultraviolet points surrounded by soft glowing clouds. And then, as I was watching, a different galaxy, a million galactic diameters away, began to light up with stars. There. Then another. Another. Billions of galaxies were sparkling with stars.

Look, Uncle, I said. Do you see what has happened in our universe? But, of course, Uncle Deva could not see into the universe, which was at that moment sitting quietly in Aunt Penelope's lap as she rocked in her chair. From the outside, Aalam-104729 looked the same as it had before, although it continued to get larger as always. From the outside, there was no hint of the metamorphosis inside. Come, I said, I will show you. And I compressed my aunt and uncle to small dancing dots and took them into the universe and led them through space, from one galaxy to the next, a galaxy at each stride. For billions of ticks, they said nothing. They only nodded and smiled.

I never saw anything like this, said Aunt P. I . . . I never saw anything like this. I would like some of these things in the Void. Can you bring them back home to the Void? I am not sure that is a good idea, dear, said Uncle Deva. Why is it not a good idea? Because, said Uncle, the Void has its own essence. And this universe has its own essence. These beautiful lights should stay here, where they belong. We can visit. Oh, it is such a beautiful, beautiful thing you have created, Nephew. I did nothing but make a few organizational principles, I said. I am tired of your modesty, said Aunt P. For once, can you not admit that you are a genius? You are an artist of the highest rank. And a mathematician of the highest rank, said Uncle Deva. And a physicist. All of those things, said my aunt. We have genius in our family, I have always known that. Wait, said Uncle Deva. I can hear something. Listen. Do you hear it? I hear it, said Aunt P. It is music, lovely, like in the Void, but different. It is the music of the galaxies, said my uncle.

Planets

While Uncle Deva, Aunt Penelope, and I were strid-
ing between the galaxies, the largest stars underwent
yet another transformation. With almost all of their
hydrogen fused into helium, these stars could no longer generate sufficient heat to fight against the inward
pull of gravity, and they began contracting again.
The collapse picked up speed. As the gaseous material crashed inwards upon itself, the center of each
star grew denser and denser, and the temperatures
mounted and mounted to a much greater degree than
before. Eventually, the temperature was so high that
the helium atoms began fusing together, to make beryllium. Then the beryllium atoms, too, began fusing to
form boron and carbon and oxygen. As the collapse
continued, the temperatures increased still more,
and heavier and heavier chemical elements were syn-
thesized: fluorine and neon, sodium and magnesium,
aluminum, silicon, phosphorous, gallium, yttrium,

molybdenum, palladium, cesium, barium, tungsten and osmium and iridium and radium. All of these complex atoms I had contemplated as theoretical possibilities, but it was a pleasure to see them actually created by inevitable events at the centers of stars, without any intervention by me. Cause and effect, cause and effect, cause and effect. On and on and on, the atomic nuclei rushed into one another and joined and produced heavier and heavier nuclei, bigger and bigger atoms. The energy release was enormous. Of course, we all understood—Aunt P and Uncle D and I, as we watched in amazement—that no material entity could withstand such colossal energy without breaking apart. Sure enough, each star soon exploded, spewing all the new chemical elements it had manufactured into space. Each detonation reverberated far into the surrounding nebulae, hurling out fragments of new matter and flaming with the luminosity of an entire galaxy. The cosmos flared and crackled and boomed, there were billions upon billions of such explosions. One by one, the beautiful stars were destroying themselves, starting with the most massive stars. Each left behind a rapidly spinning dark core and a cloud of faintly glowing debris.

At this juncture—I was counting the ticks of the atomic clocks, and it was now about 2×10^{31} ticks since the formation of the first atoms—the material that floated about in the galaxies consisted not only of hydrogen gas, but also of lumps of iron and carbon and silicon and other chemical elements forged in the stars. This enriched material did not so much float as it swirled in eddies, here and there, imparted with rotational motions from the lopsided explosions of the stars. In each galaxy, many beaconing lights remained, the lower-mass stars that had not exploded but were quietly burning down to dim nubs.

But the cosmos was not quiet for long. Under the relentless force of gravity, the cold swirling matter again began to pull itself together, to compactify, and to collapse. Soon the enriched material had again formed gaseous spheres, which were growing hotter and hotter as they contracted. In another 10^{31} atomic ticks or so, the gaseous spheres formed a second generation of stars. I must say that whereas I had been oblivious to time before the first clocks, now I found myself almost obsessed with time. Against my better judgment, I kept looking at the hydrogen clocks, ruminating on how much time passed between events. Another 10^{31} atomic ticks. Then another.

These new stars were different from their forebears. First, they contained a mélange of chemical elements and not simply pure hydrogen. Second, each of these stars was surrounded by a revolving disk of gas and debris, accented here and there with lumps of solid, compacted material. The lumps, however, were too small to ignite nuclear reactions at their centers. Instead, they collapsed to form inert solid balls. Having condensed out of a rotating disk, revolving about a central star, these solid balls also orbited the central star. These were the first planets. Planets. As with stars, I was fascinated by the effortless appearance of such distinct objects in space. In fact, planets orbited the majority of second-generation stars. There were far more planets than stars. Some solar systems contained only a single planet orbiting its central star, like a hydrogen atom. Other stars harbored as many as a hundred planets.

And what an extraordinary assortment of planets they were! Some were so close to their central star that they melted into spheres of molten sulfur and silicon and iron. Others were sufficiently far that they froze. Some were at intermediate distances, such that atoms and molecules covered their surfaces in liquid oceans, neither too hot to evaporate nor too cold to freeze. For example, there were planets of liquid water, of liquid ammonia and methane, of liquid bromine, of liquid mercury. The liquid oceans were particularly beautiful. Jostled by winds, their surfaces rippled with liquid waves. These glided across the surfaces, crest to trough to crest, glittering with starlight and reflecting the colored atmospheres above. Some liquid waves were so delicate and slight that they dissipated after traveling a short distance, barely leaving a memory of their presence. Others, fierce and rough, rose up to great heights and pushed a quarter way round the whole planet. I believe that the ocean waves were music in material form.

Interesting little contraptions, said Aunt Penelope, the little balls flying about the beautiful lights. What are they good for, Nephew? I don't know yet, I said. Do I have to know everything straight off? Well then, said Aunt, why did you make them? What did you have in mind? OK, OK, I said. They made themselves. Gravity made them. They're not as pretty as the lights, Aunt P said, and she reached up and flicked off a lump of potassium that had gotten tangled in her hair. Not everything has to be pretty, said Uncle Deva. These things will have some eventual use, I am sure of it. And look how nicely they whirl about at different speeds.

Following the inexorable laws of gravity, the planets closest to their central stars completed each orbit relatively quickly, while those far away required much more time. Indeed, in some extreme solar systems, an inner planet could make thousands of orbits about the central star in the time it took an outer planet to make one orbit, so that a year on the second planet would equal ten thousand years on the first. Other variations existed. Some planets were so small that they were irregularly shaped, with mountains nearly as tall as the diameter of the planet. Others were so big that they could almost ignite nuclear reactions within their interiors and become stars. Many of the new planets had magnetic fields, which looped out into space in pretty dipole patterns and funneled electrically charged particles in their vicinity.

In addition to orbiting its central star, almost every planet spun about its own axis. This spinning was again a consequence of the rotational motion of the primordial disk—which was, in turn, a consequence of the lopsided explosions of the first generation of stars. Cause and effect, cause and effect. It was almost mundane, these rigid chains of events, but the visual phenomena were far too interesting and new to be mundane. Even Aunt Penelope was amused by the spinning of the planets. From one solar system to the next, she poked at the rotating worlds, as she did with fleeting folds of the Void, until Uncle D stopped her.

The spin of the planets produced a charming effect. From any fixed vantage point on a planet, the sunlight was not constant, but varied in time. While your position faced your central star, you would be bathed in light. Somewhat later, when your planet had spun a half turn around, you would experience almost complete darkness. Thus, on each planet, a *day* of light was followed by a *night* of darkness, followed by a day, and a night, and so on, in a regular and periodic fashion. Described in other terms, the spinning of the planets naturally produced a separation of darkness and light, and this separation varied from one planet to the next. As there were billions upon billions of planets, all spinning at different rates, the length of the day varied enormously, being on some planets as short as 10^{19} atomic ticks, while on others as long as 10^{21} ticks. In short, there were trillions of different days (and nights) throughout Aalam-104729.

The progression of days and nights on the planets naturally led to regular changes in temperatures, variations in the densities of atmospheres, wind movements like cyclones and hurricanes and seaborne typhoons. But there were other, more subtle artistic effects. The slow shift of the light through each day caused shadows to drift, shorten and lengthen, producing constantly changing silhouettes. The summits of mountains, which might be pink in the mornings, turned violet and amaranth in late afternoon. At certain times of the day a landscape might appear craggy and hard, and at other times the same landscape could seem delicate and soft, like evanescent veils in the Void. These phenomena could not be quantified, like temperatures and densities. Instead, they heightened one's sensations. They seeped into one's insides.

Mr g

Like music, they created a feeling that was not there before. They absorbed and reshaped the world of the imagination. With changes in light, shapes constantly changed. Air sparkled and glowed, then subsided to near invisibility. On the planets with volatile liquids, great clouds of water or ammonia evaporated into the sky, and these produced further variations in light. The days and the nights yielded not only different colorations, but also different smells and sensations and tonalities of sound.

Few of these phenomena I had predicted. See there? I said to Aunt Penelope, as if it were exactly as I had intended. And there? I waited for her response. Yes, yes, she finally said, which was as close to approval as she would allow.

At a certain moment of time, a particular planet in the universe completed its first rotation, before any other planet, the end of its first day. This was the first day in the universe. I noted when this happened, and it was good (or at least satisfying), and this was the end of the first day on that planet. Then, in another galaxy, 10^{29} light-atomic-ticks away, another planet completed its first rotation, its first day, and I noted when this happened, and it also was good, and this was the end of the first day on that planet. Then, 10^{30} light-atomic-ticks away, another planet completed its first rotation, and then another, and another. At various places and times in the universe, various planets, all with different rates of rotation, completed their first days. There, and then there, and later over there. There were billions and trillions of first days, all of them good. All in all, I was satisfied with what I had done.

The
Emptiness
in
Somethingness

As I glided through the cosmos, I was taken with the relatively vast distances between things. Even though there was matter, the great majority of space was almost completely empty—not empty in the fashion of the Void, but possessing extremely little material. Galaxies of stars and planets and other material filled only about one-tenth of 1 percent of the volume of space. The other 99.9 percent of the universe was almost complete vacuum. Even within the galaxies, solar systems were far apart from one another. Starting at one solar system, I often had to travel a distance equal to the size of ten thousand solar systems to get to the next solar system. If intelligent beings ever arose on a planet in Aalam-104729, they would be separated by huge distances from other planets with life and probably never know of one another. And the separations would grow only larger with time, as the universe continued to expand.

Dissatisfactions, Disagreements, and Other Unpleasantries

After watching the formation of galaxies and stars and planets, I had a sensation unlike any I've ever had before. It was a kind of fullness. But it was more than a fullness, it was an *overfullness*, because I felt as if new things had been created within Me—an odd turn of events, since it was I who had created Aalam-104729. Or more precisely, I had created the laws and organizational principles, the matter and energy, from which everything followed. One might have thought that every new thing in that universe was already inside Me. But this did not seem to be the case. As with the invention of the quantum, I felt that I had been *changed*. I felt that my imagination had been amplified and enlarged. I felt that I knew things I hadn't known before, and I felt larger than before. How was it possible that something I'd created from my own being was now larger than my being? Is it possible that the created can create its creator? I was baffled and pleased at the same time, although that pleasure eventually led to certain displeasures.

It was not only Me. All of us felt that we had been changed. Our sense of ourselves had changed. Our perceptions had changed. For example, the Void now seemed even more empty than before. The Void, of course, had always been absolutely empty of all things, a perfection of absence. For eons of unmeasured time, Uncle Deva, Aunt Penelope, and I had rejoiced in the total emptiness of the Void. That emptiness, that complete nothingness, was one of the central and eternal absolutes of existence. That nothingness was the starting point of all action and thought, in fact, was the background that defined action and thought, that defined *somethingness*. We had all felt, without articulating the feeling, that the Void might actually be necessary for our existence. As the total emptiness of the Void was clearly an essential part of its nature, we celebrated that emptiness.

Now, however, after we had all made a number of excursions into the new universe and witnessed the extraordinary material things being made, the sacred emptiness of the Void did not bring the same pleasure it had in the past. We were—if I dare say it—even *dissatisfied* with the Void. For my own part, when I moved through the Void, I now keenly noticed what was not there. Not in the abstract sense, but in the actual and material sense, as now I could compare nothingness to atoms and electrons, to spiral galaxies and long trails of luminescent gas, to stars exploding and spewing their elements into space. I could not help but feel a bit disappointed in the *plainness* of our habitat in the Void.

Aunt Penelope did not take the same delight in gathering up little pieces of the Void for her personal preoccupations. What's this? she whined during a recent outing as we strolled through the Void. You know what that is, said Uncle Deva. It is a scrap of emptiness that you will take back home and put to some good use. But it is nothing! said Aunt P. Yes, said Uncle, it is precisely nothing. It is a nice piece of nothingness. Perhaps you can make a dress from it. No, said my aunt. I will not. It is truly nothing. It is really nothing. I want to make a dress from the galaxies and stars. Oh, what a magnificent dress that would be! I would shine, and everyone would want a dress like mine. Nephew, my aunt said to me, can you please be a good nephew and bring back some material from that universe. I do not need a lot. Uncle looked at me with disapproval and annoyance. Do not butt into this, Deva! said Aunt P. This doesn't concern you. The material in the universe should stay in the universe, said Uncle Deva. Everything has its own place. Don't be so self-righteous, said Aunt P. Just a little while ago, if I remember correctly, you were saying that you would like to look at a few mountains now and then on our walks through the Void. Don't deny it. You always exaggerate, said Uncle. I asked for only *one* mountain. Not mountains plural. All right, one mountain, said Aunt P. You admit it. So why can't I have a few galaxies and stars if you can have a mountain? What do you say, Nephew? Can you see your way clear to bringing back one mountain for your uncle and a few galaxies for me?

I refused to get caught in the middle of squabbles between my aunt and uncle. Let me think about this, I said. I'm not sure if . . . You are always thinking, said Aunt P. You think about this, and you think about that, and then you think some more about this. Why can't you just *do* it. Go fetch me some galaxies. I want to make a dress. I am tired of this nothingness here. Tired, tired, tired. All we have is a bunch of nothing here. I want *something*. You shouldn't order Him about, said my uncle. Deva, I've had quite enough of your butting in, said Aunt P. You are beginning to tire me yourself, just like the Void. You are empty. You are full of nothing. If you had a bit more ambition, you could have . . . You're going too far, said Uncle. You've gotten yourself into one of your states. No I have not, said Aunt P. I'm just beginning to see things as they really are. I don't like the way you are acting right now, said Uncle. You need to calm down. Don't tell me to calm down, said Aunt. That is condescending. Do not condescend to me. My uncle reached out in an attempt to caress Aunt. Don't get near me, said my aunt. And don't expect to sleep with me for a long long while. Don't flatter yourself, said Uncle. I don't know who would want to sleep with a nag like you.

Please, I interjected. Don't fight with each other. He started it, said Aunt Penelope. At that, both Aunt Penelope and Uncle Deva went stomping off in different directions. I never followed them in these kinds of altercations, preferring to let them wander about on their own and recollect themselves. I watched them as they went away, becoming fainter and fainter as they slipped behind accumulating layers of nothingness. Finally, they both disappeared. Presently, the Void grew calm again and began to resonate with soft music.

Aalam-104729 had been left on a gentle outcropping of nothingness, not far away, and it was expanding as always. The universe was off to a good start, with galaxies, stars, and planets, and I found myself wondering what other objects I could make. I wanted a lot more matter, a lot more energy, a lot more everything. The one universe was very nice, but as I stood looking at it now, it seemed rather small. Other potential universes were flying about the Void, throbbing and spinning but empty. Some of them might become far grander than Aalam-104729. What wonderful new things might I fill them with! All I had to do was decree a few more organizational principles, specify a few parameters, and they too would burst forth with matter. I wanted to make galaxies a hundred times larger than the ones I had seen in Aalam-104729. I wanted to make stars as big as galaxies, planets as big as stars, solid oceans. And I just wanted more.

At that moment, there were at least 10^{189} empty universes careering through the Void, all beckoning with their possibilities and potentials. I reached up for one as Aunt Penelope had done. I will start with this cosmos, I thought to myself. It was a fat spheroid, not silky on the outside as some of the others but mottled and tough. This one has ambition, I thought. It will challenge me. As I prepared to enter it, Aunt Penelope called out to me from wherever. What are you doing, Nephew? I was going to begin working on another universe, I said. What for? said Aunt P. She appeared in the distance and marched towards me at a brisk clip. I wanted to try something bigger, I answered.

And better. Aren't you happy with the universe you've made? asked my aunt. Yes, but . . . You haven't finished it yet. Yes, but . . . Nephew, you are impatient. Didn't we talk about that before? You are too much in a hurry. You will not do good work that way. And, if I might say so without giving offense—we are family after all, and one should be able to say these things to family—you are acting *greedy*. Plain greedy, and it does not become you.

I was stung by Aunt P's remarks. Too often, she found fault with something I did, or frowned at me in that unpleasant way of hers, or simply woke up on the wrong side of something or other. She was not entitled to speak to me that way, or to Uncle D for that matter. For eons of time, she had been walking all over Uncle, treating him as worth less than nothing, and he had just accepted it, hardly fighting back. But it diminished both of them. Greedy! How was I greedy? What was the harm of wanting to fill up a few more universes? In my opinion, Aunt P was off base, way off base. And why had she said such a mean-spirited and hurtful thing to me? She was compensating for something, something lacking in herself. Well, her barbed comments were not worth a reply. I was not about to lower myself. Who did she think she was talking to?

I went for a long walk in the Void. I am not sure what I was thinking, but I wanted to be alone. Time passed.

What did it matter how much time passed, anyway? Time passed. I traveled great distances. I went this way and that, scarcely noticing the hills and the valleys of nothingness, the folds upon folds of emptiness, the utter vacuum. I am not sure what I was thinking, or how much time elapsed. It might have been eons. I reminisced about epochs past, before the invention of time, when Aunt Penelope and Uncle Deva and I all spoke at once. None of us could hear the other, as we were all talking on top of one another, but it was part of how we related, and there was a certain pleasantness and familiarity to it. I reminisced about potential thoughts I had had, long before I decided to create anything, when even the thought of creating something was only a potentiality, a possibility. How sleepy we all were. I marveled at how Aunt P and Uncle D had changed during the infinite time I had known them, but especially in recent epochs. They seemed to be growing closer to each other in some ways, keeping their distance in others. And I thought of my own changing sensitivities, how I had been aware of the ubiquitous music filling the Void only after the creation of time. Before then, music, happening all at once, seemed just another aspect of existence, like the nature of thought. So much had happened in a relatively short period of time, certainly short compared to the unending sprawl of existence before. I walked and walked and walked, huge distances in the Void, but huge distances in the Void are infinitesimal compared to the infinite. Eons passed.

When I came back to where I had started, there was Aunt Penelope, exactly where she had been before, as if not even a single atomic hydrogen tick had gone by. And I realized that she had been correct. I had been greedy. In the past infinity of time, I had never known myself to be greedy, but then again I had never had anything to be greedy about. Matter was a recent invention. I had been greedy. I felt embarrassed. Immediately, I let go of the fat, empty spheroid with the mottled exterior, and it flew away in haste, joining the myriad other empty universes zipping about. I'm sorry, I said to my aunt. You are right. I was acting greedy. One universe at a time. We will see this one through. Thank you, said Aunt P. One of your admirable qualities, Nephew, is that you admit your mistakes. Not like certain other parties who will go unnamed.

The Origins
of Life

After my humbling conversation with Aunt Penelope, I decided to attend again to Aalam-104729. Since my last visit, a number of fascinating changes had occurred. As a result of the nuclear reactions in the first generation of stars, the most abundant elements in the universe were hydrogen, helium, oxygen, and carbon, so I expected many molecules to have formed from these elements. And I was right. Water, consisting of two atoms of hydrogen bound to one atom of oxygen, was plentiful in at least one planet in every dozen solar systems—covering its surface in liquid oceans and floating above in gaseous vapors. Another common molecule in planetary atmospheres was methane, comprised of one atom of carbon bound to four atoms of hydrogen. And carbon dioxide. And ammonia. Sunlight filtered through these atmospheric gases in a lovely way, causing the air on certain planets to glow in crimson, turquoise, and cadmium yellow.

Other phenomena were less expected. The atmospheres of a great many planets were repeatedly cracked by jagged bolts of electricity. These spectacular discharges of electricity slammed energy into the primordial atmospheres and formed complex new molecules: sugars and carbohydrates and fats; amino acids and nucleotides.

Among all the atoms, carbon was supreme at bonding with other atoms. It had four electrons available for pairing, the maximum number for the smaller atoms. As a result, carbon atoms could link together and gather up hundreds of atoms of oxygen, hydrogen, and other elements in long, gangly chains. Or, instead, form hexagonal rings and other elaborate structures. Nitrogen, able to share three electrons with other atoms, was also excellent at bonding. Other elements had different bonding abilities, leading to many patterns. Atoms in molecules formed linear chains, planar triads, tetrahedrons, octahedrons—some folding upon themselves in the most marvelous ways. All of which was caused by the particular electrical attractions and repulsions between atoms, arising, in turn, from the precise orbits of their electrons. These orbits, finally, were rigidly prescribed by the laws of the quantum.

As with the planets and stars, I had nothing to do with the manufacture of these molecules. They formed by themselves, irresistibly following the creation of matter and the small number of principles with which the universe began. Cause and effect, cause and effect. I was a mere spectator. But I would watch over the progression of events, as Aunt P had suggested, and intervene if things began going awry.

As soon as large molecules were formed in the planetary atmospheres, they plummeted down through the air and sank into the oceans. And dissolved. Just as carbon is the atom most suitable for forming complex molecules, water is the superior liquid for dissolving other molecules. Owing to the positions of its electrons, water molecules can gently pull apart other molecules, attach themselves by electrical attraction, and escort the guest molecules as they swim about.

In time, the oceans of quite a few planets became a thick hodgepodge of carbon and nitrogen-based molecules, water, and fragments of other molecules. These bits and pieces proceeded to collide with one another at great frequency as they moved and jostled about in the warm seas. Even on a single planet, there were trillions upon trillions of such molecular collisions every tick of the hydrogen clock. With so many encounters, all kinds of new things occurred. New molecules were created. Some molecules stuck together to form bigger molecules. Some rearranged or tore off pieces of each other. Some extracted energy from other molecules by plucking off their electrons. Various architectural structures formed, such as spherical cavities or solid ellipsoids, held together for a few moments, and then came apart. It was trial and error, trial and error, trial and error. It was trillions of scientific experiments performed every atomic tick. I could hardly wait to see what would happen.

One of the molecules, a long chain of carbon, hydrogen, oxygen, phosphorous, and nitrogen atoms, had the ability to replicate itself. At each section of the chain, the electrical attractions were just right to snatch a matching section from the jumble of raw material floating around it and duplicate itself. This master molecule could do more than replicate. It also served as an intermediary in assembling other molecules. In action, the molecule seemed almost purposeful, yet it was dumb lifeless matter, like the rest.

Then a curious thing happened. Quite by accident, in the quadrillions of new structures that formed every tick, one of the self-replicating molecules found itself lodged within a closed cavity of other molecules. The wall of this cavity, only a couple of molecules thick, encircled its own little world. And how small it was, trillions of times smaller than a planet. Yet this tiny cellular world had a certain wholeness, an outside and an inside. Outside was the thick ocean, full of sugar and carbohydrates and amino acids. Inside was one of the replicating molecules and other carbon and nitrogen-based molecules that had come along by chance. The cellular wall allowed some molecules from the outside to enter. Others were refused. However, even such cells could not sustain themselves if they did not have a source of energy. Energy was critical. In the Void, Uncle Deva, Aunt Penelope, and I had an infinite supply of energy. But here, in Aalam-104729, energy was a limited commodity. There was only so much and not more, and one had to find energy as best one could to maintain and survive.

It was only a matter of time before some of the cells had, again quite by accident, amassed all of the ingredients to preserve themselves indefinitely: the ability to procure energy by disassembling sugars, which harbored a great deal of electrical energy in the repulsive force between their electrons; the ability to reinforce themselves with supplies through a selective cellular wall; and the ability to reproduce themselves by enclosing a replicating molecule. Such cells formed in the oceans of many planets. They thrived on the rich sugars and other molecules floating in the warm seas around them. They exchanged materials with the outside world. They grew. Then they replicated their insides, split apart, and doubled their numbers.

Were such things alive? It depends on what one means by life. They were organized. They responded to their environment. Unlike mountains and oceans, they could grow and reproduce themselves. But in other, more essential ways, they were just dumb material. All of their wonderful mechanisms happened without any thought. In fact, there was nothing resembling thought within the sparse and limited protoplasm of their bodies. They could not communicate. They could not originate ideas. They could not make decisions. They certainly had no self-awareness. What few electrical impulses surged within them served solely to maintain and preserve themselves, and even these occurred quite automatically, like a standing rock that falls and topples another nearby rock, which falls and topples another rock, which falls and topples another, and so on. No matter how many rocks you have in such a progression, would you say that the thing has any thinking capacity? Certainly not. The rocks are just dumbly obeying the laws of gravity. So, while I was amused by these self-replicating cells, I would not say they were alive in any meaningful sense of the word. I would call them fancy inanimate matter. That's what I would call them—fancy inanimate matter. And I was in no hurry to make animate matter.

I went back to the Void and gave a full report to Aunt Penelope. By this time, she had made up with Uncle Deva and was even allowing him to brush her hair. I found the two of them together, she sitting contentedly in her chair and he standing behind her. Now you've got it right, she said. Just there, just like that. There. That's it. That's it. Now you've got it right.

I was wondering when you would return, said Aunt P when she looked up and saw me. I should check up on the thing now and then if I were you. All is going well, I replied. Do not get smug, Nephew, she said, and threw a look at Uncle, as if challenging him to take issue with her sharp comment to me. But he said nothing and continued to brush her hair in long, rhythmic strokes.

It was a while before my next visit to Aalam-104729. To be exact, it was 2.5×10^{32} atomic ticks, but I wasn't watching the time. I had been loosely following events in the new universe, but only loosely. Now that I gave it my full attention again, I was astonished at what I discovered. The fancy inanimate cells had continued to evolve, constantly buffeted and altered as they were by the molecules floating around them. Apparently, a large number of additional molecular possibilities had been explored, again all by mindless chance. Some of the cells had manufactured molecules that could directly use sunlight for energy, converting water and carbon dioxide to sugars. The by-products of these new chemical reactions included oxygen gas, which bubbled out of the water and into the atmosphere.

Oxygen, in gaseous form, is caustic. It burns. It corrodes. It snatches electrons from other atoms and destroys them. A great many of the light-utilizing cells were annihilated by their own productions. But some had oxygen-resistant skins and had even evolved into new cells that could use oxygen to extract energy from sugars and fats.

On some planets, the new oxygen-utilizing cells had banded together to form larger and more complex organisms. These larger things, made of millions and billions of cells, continued to change as the single cells had done. As new molecular possibilities occurred, always by trial and error, the cells in these composite organisms did not all evolve in the same way. Some of the cells took on specialized functions, such as processing waste material or circulating the needed oxygen or providing mechanical means for the organism to move more easily about. Some of the cells even evolved to coordinate and control the activities of the other specialized cells.

All this had happened in my absence! Mindlessly following the rules of chance and necessity, the warm seas of the planets were churning out highly organized and efficient multicellular organisms. I felt a slight embarrassment that so much could proceed without any direction by me.

At this point, I was hesitant to call these things

entirely inanimate. And I could see the rudiments of brains. Not brains with ideas, but conglomerations of cells that were clearly coordinating other cells. Undoubtedly, these specialized conglomerations would become more and more complex as the organism became more complex. The coordinating and control cells would send more electrical signals to one another. They would develop feedback loops. They would have a sensation of changing in response to stimuli. At times, they might exchange signals between themselves with no survival value but simply as a statement of shared existence. I could see the trend. Eventually, the things would have some kind of recognition that they were independent entities, separate from the external world. They would perceive themselves from outside of themselves. In short, they would become *aware of themselves*. And then they would *think*. It was only a matter of time.

How mistaken I had been. To believe that I could purposefully decide whether to create animate matter or not. As was now apparent to me, animate matter was an inevitable consequence of a universe with matter and energy and a few initial parameters of the proper sort. If I wanted, I could destroy life. But I was only a spectator in its creation.

I was surprised. I was moved. I was concerned. What was this thing that had been set into motion?

Mr g

First there was time. Then space and energy. Then matter. And now the possibility of life, of other minds. What would the new minds think? What would they grasp? Hadn't I wanted this? Yes, I had wanted it. But also, I had not wanted it. Certainly I wasn't prepared. I could feel the weight of the future, heavy, bristling with possibilities. But I could not see the future. The future was a dim, throbbing thing, an invisible galaxy. Was it out of control, *my* control?

Free Will?

On occasion, I go into meditative states that last an indefinite period of time. Before the first atomic clocks, these states were truly of incalculable duration. Aunt Penelope might tell me, after I had emerged from one of my meditations, that I had been gone a very long time, that she and Uncle had completed many sleepings and wakings while I was absent, that the music had practically ceased in the Void, and so forth. But to me, no time had passed. Or rather, no events had passed worth my attention. Even with the invention of devices to measure time, I still supported the view that time had meaning only in its relation to events. If no events occurred, or events of no significance, then one could say with some justification that no time had elapsed. At any rate, that was how I felt after each of my meditations in past eons. Indeed, the purpose of such meditations, for me, was to free myself from events, to bring my mind back to itself, to transform

myself and my thoughts into a state of pure, instantaneous being. Certainly that was easily attained in the
Void. In the emptiness of the Void, events never happened except for the occasional excursion by Uncle
Deva or Aunt Penelope—and, with due respect to my
uncle and aunt, those events could hardly be called
significant.

Now, however, things had changed. With so many
happenings in the new universe—indeed the new universe was a shouting match of events, one on top of the
other—there was never a moment free of significant
events. Even in the Void, that most perfect and absolute condition of nothingness, one was always aware
that events were racing forward pell-mell in that small
but ever-increasing sphere. No matter what one was
doing in the Void, no matter where one was in the
Void, one could *feel* the crowd of new developments
in Aalam-104729. One could feel the present rushing
headlong into the future, even though that future was
faint and fraught with uncertainties. The trillions of
other universes flying about, throbbing with potentialities, seemed as nothing compared to the explosions
taking place in Aalam-104729. No longer could I meditate in complete stillness.

Thus it was that I had removed myself from existence, while at the same time being aware of the
clamorings within the new universe, when Belhor and

Baphomet appeared again. Belhor looked as he had on the previous visit, a tall and thin figure, dignified and dark. Baphomet, still squat and ugly with his interminable smirk, seemed to have developed a swagger. And now there were two Baphomets. Standing behind the original was a second, smaller Baphomet, who cowered in the most pathetic way and constantly bowed to everything in sight.

"I hope I am not disturbing you," Belhor said to me.

"We hope we are not disturbing you," said the larger Baphomet. "We would never ever want to disturb such an eminence as yourself. Not ever."

"No, we would not," said the smaller Baphomet. At which point the larger Baphomet jerked around and kicked the smaller beast, who let out a pitiful yelp. "Quiet," said Baphomet the Larger. "I will tell you when you can speak." "Yes, master," said the smaller Baphomet. The larger Baphomet did a backwards somersault and grinned.

Of course they were disturbing me. I was triply disturbed. But I believe in some consideration to others, regardless of their actions, so I merely replied, "What do you want?"

"We have much to discuss," said Belhor. "Many things have happened since we last met. Interesting things, I would say. It seems that primitive life has arrived in our little universe. Inevitably, the primitive life will evolve and become more . . . shall we say, complicated. Self-awareness will come next. And intelligence. It is only a matter of the passage of time. Would you agree?"

"Yes," I said. "Unless I intervene. You are well informed."

"I make it my business to be informed," said Belhor.

"Oh, my master very definitely makes it his business to be informed," said Baphomet the Larger. "My master knows everything about everything. Isn't it grand." Belhor gave Baphomet the Larger a fierce look, at which point Baphomet turned and kicked Baphomet the Smaller.

"My request," Belhor said to me, "is simply that you *not* intervene. Allow these primitive creatures to evolve and develop self-awareness and intelligence."

"I will consider your request," I said.

"With intelligence," said Belhor, "the new creatures will at least have the *impression* of making decisions on their own. Of course, whether they know it or not, *we* know that they will be following the same laws and rules as inanimate matter, the laws and rules that you have laid down. Their behaviors and actions will still be fully prescribed in advance, aside from the slight modifications arising from your quantum concoction. But the creatures will have the impression of freedom of choice. We can let them have that impression, can we not? What harm would there be?"

"Do not play games with me," I said.

"I would never play games with you," said Belhor. "I do not play games with anyone, and especially not with you. I am not asking that you do something. What I am asking is that you *not* do something, that you not intervene, that you let matters follow their own course."

"Yes," I said, "I understand what you are asking. I will consider it. There are pros and cons to the development of intelligent life in the universe, and they must be carefully considered. At the moment, I do not see any harm, as you say, in allowing the new animate life-forms to have the impression of making independent decisions on their own."

"Good," Belhor said. Slowly, he moved to where Aalam-104729 lay on its side and picked it up. The universe made a small, moaning sound. "It is so precious," said Belhor, "with so many possibilities. Indeed with an infinite number of possibilities. If I may, do you remember our previous conversation, in which we agreed that with a sufficiently large number of outcomes, it would be impossible to imagine everything that might transpire in the future? Do you remember that conversation?"

"Yes, I remember it. I do not forget anything."

"No, of course not," said Belhor. "I just wanted to make reference to that conversation. Please, with your permission, let me continue. I want to discuss a matter of principle with you. We have recognized that our intelligent creatures will have the impression of making their own decisions, while at the same time being totally subject to your rules and laws. But the question now is: Will you have *foreknowledge* of their decisions and actions, even if their atoms and molecules are following all of the rules that you have laid down? And even if you could in principle have foreknowledge, would you be willing to relinquish that foreknowledge in some cases? Please hear me out. Many of these creatures will have brains. There are an extremely large number of possible arrangements even for a small brain. Consider a typical atom, like carbon. Say, for example, that it has 20 possible configurations. There might be 10^{14} atoms in a single one of your new cells, so that there are then $20^{10^{14}}$ different possible configurations of a single cell. In one rather modest brain, with 10^{12} cells, there would then be $20^{10^{26}}$ possible configurations. That number is enormously larger than the total number of atoms in a galaxy. As you can see, there is a staggering number of different configurations in even one modest brain, all of which may have an impact on a decision being made by that brain."

"What of it?" I said. "I can do that calculation." The more I knew of Belhor, the more impressed I was with him. And concerned.

"Of course you can do the calculations," said Belhor. "To be frank, it is a pleasure for me to converse with someone of your intelligence. And I hope the same for you. Of course you can do the calculations. But why would you want to? At each moment, there is such a huge number of possible configurations for a single modest brain. Now consider that there will be billions of brains on each planet, and billions upon billions of planets. Why would you want to keep track of all of those brains, all of those tedious calculations, all of those possibilities? And remember that the displacement of a single atom in any of those brains might change the outcome of a long sequence of events, ending in a different decision or action."

"You are clever," I said. "But I do not understand what you are getting at."

"I believe that you do," said Belhor. He was looking at me intently now. Even face on, he was thin, so very thin, like the sharp edge of something. "I am saying that, just as you should not intervene in the development of intelligent minds in our new universe, I request that you should not attempt to predict the behavior of those minds. Let them make decisions and take action without your foreknowledge. The creatures are hardly worth troubling yourself about. They will still follow your rules and laws. But there are so many possibilities. Let the creatures act without your foreknowledge. They will have the sensation of making their own decisions . . . in fact, more than the sensation. But they will still be following your rules. Again, I am not asking you to do something. I am asking you not to do something."

"If these hypothetical beings—and I have still not decided whether I will allow them to exist—if these hypothetical beings make decisions without my fore-knowledge, then they will not be within my control."

Belhor said nothing. He continued looking at me.

"Oh, I think the Big Guy is worried," said Baphomet the Larger. "You know, I never thought the Big Guy worried about anything. My master has got Him worried."

"Their atoms and molecules will still be following your rules," Belhor said to me.

"Yes," I said, "but with the large number of possible configurations, as you have pointed out, small distur-bances could change outcomes. It will require some effort to predict all outcomes in advance."

"Precisely," said Belhor. "Do you insist upon hav-ing complete control of everything you have created? We have discussed this matter before."

"I must consider all of this," I said. In my own mind, I was thinking that Belhor was quite right, although I would not give him the satisfaction of immediately approving his request. I did not really want to calcu-late the zillions of possible configurations involved with each decision of each intelligent creature on each planet.

"Then you will consider it?" said Belhor.

"Yes, I will consider it."

"Good," said Belhor, and he smiled in that unsettling way that he had once before. "Now, I am interested again."

"We are all interested," said Baphomet the Larger. "Very interested."

Belhor bowed. "I have enjoyed our conversation immensely. All of us here in these regions have responsibilities towards the new things you've made. To life. To life."

Goodness in Every Atom

I did not tell my aunt and uncle about Belhor et al.'s latest visit, as I knew their hostility towards him. Nonetheless, Uncle Deva seemed uncannily aware of the conversation. Or, if not the conversation itself, at least the issues surrounding life in the new universe. During one of my recreational strolls through the Void, Uncle cornered me. Aunt Penelope was napping, he said, and it would give us a chance to chat "without impediments."

So, Uncle said cheerfully, I gather that we will soon have animate creatures hopping about in our universe. A very good development, if I may say so. I hope that you have not been thinking otherwise. He made this last remark in a casual tone, but I knew that he was slyly probing me. The easiest thing would be to let it happen, I said. But I have not completely decided on the matter. Oh dear me, said Uncle. Uncle

Deva never chastised me in the manner that Aunt P did, but from eons of living with him I knew what "Oh dear me" meant. He was vexed with me. You have done such a masterful job, said my uncle. The galaxies. The stars. The what-you-may-call-its. Even the solitary little hydrogen atoms floating around. But surely the grandest accomplishment of all would be the creation of intelligent life. There are problems, I said. Yes, of course there will be problems, said Uncle. But doesn't everything have problems? Nephew, what meaning does your universe have without other minds in it? It has beauty, I said. Yes, it has beauty, said my uncle. But who is there to appreciate the beauty, aside from you and me and your aunt? Wouldn't the beauty have more meaning with other minds to admire it? Wouldn't it be *transformed* by other minds? I'm not talking about a passive admiration of beauty, but a participation in that beauty, in which everyone is enlarged. We three are not of the same essence as the universe. But living creatures born into that universe, made of the same material, are of that essence. You have told me yourself that the life-forms are made of the same atoms as everything else in the universe. The beauty you speak of—the stars and the oceans and so forth—is part of *their* beauty, those living things. And so much enhanced by their participation, by their absorption of that beauty and then the responsive outflowing of their own beauty. It is a spiritual thing, don't you see?

I so loved Uncle Deva. He was sincere in his beliefs, and sweet. Don't you want your universe to have some recognition of itself? continued my uncle. I mean, the minds within it? As beautiful as it is, a mountain cannot have recognition of itself. Don't you want some bits of your universe to know that they are part of a whole, part of a pattern, that some glorious act created time and space and matter and set the whole thing in motion? It was not so glorious, I said. As you remember, I was tired of the unending nothingness followed by more nothingness. I wanted a change. I wanted somethingness. That's all.

You can say what you want, said Uncle. But even if you had no grand purpose in mind . . . the fact is, the creation of the thing was glorious. An act can be glorious whatever its intention and purpose. Intelligence, awareness, mindfulness are going to connect the pieces of our universe in a way that inanimate matter never could.

Uncle looked at me affectionately and sighed. I can see you are troubled, he said. What are you concerned about? I am concerned that something unpleasant will happen, I said, something terrible. Maybe many terrible things. If an act can be glorious whatever its intention, an act can also be disastrous whatever its intention. I am concerned that the intelligent beings in the new universe will come to some harm, that they will suffer. Are you still thinking about Mr. Belhor? asked Uncle. What does he know? Your *goodness* will prevent suffering. You must believe that. The animate life, once it has developed intelligence, will feel your goodness. No suffering can come from that. I am not so sure, I said. Have faith, said Uncle. Your goodness fills every atom in the universe. It will flow into every creature created. Suffering cannot occur in such a cosmos.

I wish I were as certain as you, I said. I feel this rush, this vibration going through me, the future. The future is happening.

Then it is settled, said Uncle Deva. I have been waiting for this moment for time upon time. Yes, the future is happening. Your aunt and I have been discussing this possibility for some time, and we have a few things to suggest, just a few things, about what we want our creatures to look like.

Bodies and Minds

But it didn't matter what Aunt Penelope and Uncle Deva wanted the new creatures to look like. Because that development, like almost everything else in the new universe, happened on its own by trial and error, with no need for meddling by outside parties.

In the trillions upon trillions of galaxies, and the billions of planets in each galaxy, every imaginable form of life arose. The light-utilizing creatures developed into wondrous vegetations, some tall and skinny and deeply fixed in the planetary soil, others small and delicate and gorgeously colored. They were rough and barky, soft and silky, sticky, moist, dry, gelatinous, sharp, rounded, generous and open, closed and tight as if protecting a secret. Some lived on land, some under the oceans. Some floated in the air, blown about by winds. Some even left their home planets altogether and drifted through space, finding raw materials in the long wisps of interstellar gas. Some were heavy with exterior scales and bark, some so slight as to be almost invisible, barely a few molecules thick. And so many shapes: circles and disks, spirals, fans, sponges and sheets, floppy flats, filigreed meshes, thick blobs and windings. Generally, the various vegetations lacked their own locomotion. However, nourished by sunlight as they were, they all found habitats and positions where they could orient themselves towards their central star. Their molecular machinery turned sunlight, water, and carbon dioxide into sugars, and they lived on those sugars.

The oxygen-utilizing creatures were more complex, with their enhanced metabolisms and efficiencies. They had more intricate organs. They moved about from one place to another. They fidgeted and grasped. They ate. They reshaped their surroundings. A great many of these animals remained in the liquid oceans where they first formed and developed streamlined bodies so that they could swish through the water with minimal friction, propelling themselves forward by wriggling their smooth surfaces. Others grew feathery appendages, which they flapped against the air, creating sufficient upward force to counteract gravity. These feathered animals flew through the atmosphere in graceful swoops and glides. Some of the oxygen animals developed large sacs of low-density gases so that they could float about in the air. They steered by emitting little jets of fluid and gas. Creatures that took to the solid land moved about on two or more appendages, some with as many as a hundred, which they thrust back and forth in jerky movements.

To aid in responding to external stimuli, many sensory devices evolved: electromagnetic sensors, acoustic and vibrational sensors, thermal sensors, molecular sensors. Specialized cells, sensitive to light or mechanical pressures or particular molecules, were nestled in oddly shaped flaps and protuberances and bulges of flesh. In some star systems, advanced creatures developed only one or two electromagnetic sensors, generally placed on the tops of their bodies; in others, dozens were scattered all the way to the extremities. Some animals developed an exquisite sensitivity to magnetic fields, others to infrared radiation, still others to minuscule vibrations, with the ability to decompose slight disturbances into harmonic components and thus to create a map of the movements around them.

Anatomies varied like everything else. There were organs for processing sugars and fats and other sources of energy, organs for circulating liquids and gases, organs for disposing of wastes, organs for emitting high-frequency sounds for communication, organs for storing chemical and vibrational energy, organs for maintaining balance in a gravitational field. Oxygen animals had structural bones. They had internal electrical systems. They had multiple appendages, some bristling with sensory organs. They were sheathed in hair, fur, scales, crystals of silicon. Creatures in warmer climates evolved thin, porous skins, so that heat could be easily conducted from inside to outside their bodies. In colder climates, creatures had bulges and layers of fat just beneath the skin to hold heat in. On planets near ultraviolet stars, creatures grew thick metallic shields covering their bodies. Animals on planets with low gravity tended to be floppy and large, on planets with high gravity small and compact.

In the course of billions of planetary orbits about their central stars, billions of seasonal cycles, many possibilities were tested. Structural features that helped an animal survive were naturally perpetuated in future generations. Those features that did not were eventually eliminated, as the creature's descendants could not cope well enough with their environments to continue reproducing. As far as progeny were concerned, many of the oxygen creatures reproduced in pairs, combining their replicating molecules to make small, detached offspring rather than each adult splitting in half. In some worlds, creatures reproduced not in pairs but in triples and quadruples. These latter exchanges required awkward conjugations of bodies but allowed great variations of progenitor material.

Mr g

And the brains! As I had suspected, the masses of coordination and control cells had evolved to a fantastic degree, forming intricate networks of electrical activity. Some of these brains contained as many as ten trillion cells, each cell connecting to a thousand other cells. Over time, creatures with such brains rebuilt their environments. They made new materials and inanimate structures of their own design. Waterways. Tools. Machines. Cities. They developed advanced communication methods, such as encoding information in electromagnetic radiation or storing it in silicon-based molecules and quantum clusters. They created devices to extract energy from their central star and from passing comets. They discovered mathematics. They performed experiments. They built instruments that could sense what their bodies could not. They developed theories of the physical universe. And they discovered many of the laws and principles that governed the universe, *my* laws and principles. These mere conglomerations of atoms and molecules *discovered my laws*. And the music they made! Such music, equal to what I have created from my mind, they produced by material instruments with vibrating strings and air flows and liquid compressions. When I heard their music, from one star system to the next, I realized that these brains were participating in the beauty of the cosmos, as Uncle Deva had described.

They were aware of themselves, yes. They were thinking, yes. But they were more than thinking. They were *feeling*. They were feeling the connection of themselves to the galaxies and stars. They were grasping the beauty and depth of their existence and then expressing that experience in musical harmonies and rhythms. And in paintings. In metaphors, and words. In dance. In symbiotic transference. They imagined the cosmos beyond their own bodies. They imagined. But they could not imagine where all of it started. For all of their intelligence, there were limits to their imagination. They could not know of things that were not of their essence. They could not know of the Void. But the mystery of such things they did seem to feel, and it tingled in them and opened them up.

Time. Time fluttered and spun and wound itself up. Time stretched and compressed and dilated and dissolved. I had been mistaken about time. Although time could be measured and sliced by the beats of the hydrogen atoms, now that other minds existed time did not move on its own. Or rather, even if it moved on its own, its movement was relevant only to how it was witnessed. Time was partly conception. Time was partly a thing in the mind. Just as events. Since the universe began, nearly 10^{33} ticks of the hydrogen clocks had transpired. Stars had been born. Stars had aged, then exploded or dwindled to dim and cold ashes. Galaxies had collided. Living cells had formed. Then minds. Cities had risen on deserts. Cities had fallen. Civilizations had flourished, then ended. Then new civilizations emerged. Nothing was lasting, nothing was permanent. Living creatures, beings with minds, were the most fleeting of all. They came and went, came and went, came and went, billions upon billions of lives, each quick as one breath. Atoms converged in their special arrangements to make each precious life, held together for moments, then scattered to dull lifeless matter again.

Atom for atom, life was a rare commodity in Aalam-104729. Only one-millionth of one-billionth of 1 percent of the mass of the universe abided in living form.

Consciousness

Although it had happened quite on its own, I was fascinated to understand how *consciousness* had arisen in the new universe. What an amazing and unexpected phenomenon! You start with some dull lifeless material, you let it knock about on its own, bumped around and shaken by other dead stuff, you let it change and evolve by haphazard events, and suddenly it rears up on its hind legs and says, "Here I am. Who are you?"

Certainly I understood all the energies and forces in atoms. They were just my laws and principles. But *consciousness*—this cooperative working together of individual cells to create a sensation of wholeness, of being alive, of existence, of *I-ness*—was something else. It was a collective performance that went far beyond the individual pieces. It was strange. It was wonderful. It was almost a new form of matter. How had it happened? And how many cells did it take to make consciousness?

Mr g

I decided to do an experiment. I loved experiments. While the emergence of consciousness in the universe had required billions of planetary years, I could speed up the process to mere moments. I entered Aalam-104729. Then I scooped up a bunch of the coordination and control cells from a warm sea on one of the newly formed planets—such cells were remarkably similar from planet to planet—and I put them together on a smooth rock on the shore and let them form electrical and chemical connections between themselves. I started with a thousand cells. Nothing. Then I added more cells, a thousand at a time, until I had a hundred thousand, a tiny grey blob of material sitting on a rock. Still nothing. Just random electrical pulses, odd hums and buzzings. It was frustrating. I so much wanted the thing to wake up. I shook the rock. I jostled the grey blob. I even played a bit of music, a frisky bellantyne, hoping to nudge the thing into consciousness. I added more cells. Ten million. Still nothing. A hundred million. Not much, but the electrical activity was getting more interesting. More complex patterns were beginning to emerge, collective patterns I had not seen before. The cells were interacting with one another, but with mostly meaningless blather, like Aunt Penelope's mutterings. At this point, I had reproduced a couple billion planetary years of evolution. I doubled the number. Two hundred million. Now something unusual was happening! The gelatinous hodgepodge of cells began creating patterns of electrical activity not related to its survival. *Unnecessary* electrical activity. But not random either. The thing seemed to be reacting to itself. A little electrical peep would start in one cell and get passed around to the other cells, each of which would make its own peep, and then all the peeps would start chiming in unison, amplifying one another into a single electrical chirp. This chirp rose and fell, almost like a melody, with complex overtones. After a while, it would die down and the thing would get quiet. And then it would start up again.

Was this it? Was this consciousness? The thing I was looking for was so subtle, so delicate, and yet unmistakable once there. Clearly I did not expect my little blob of cells to speak to me—indeed consciousness evolved before speech—but at some point the mass of matter would become aware of itself.

I felt that I was on the verge. I was on the verge, and I just needed to do a bit more. I looked around for some tools to poke the thing with. There. I arranged for a twig to touch the thing ever so slightly. The cells that had been tapped shuddered and twitched and sent peeps around to the others, and they began peeping in unison until the single chirp started up. After a few moments, the little grey blob got quiet again.

Then I caused the twig to tap the thing two times, a pause, then three times. Well, this was big news. This was clearly a hello from the outside world. Now the cells that had been tapped trembled more strongly, contacted the others, and soon all two hundred million cells were excitedly chattering to one another as if they had just discovered that down was up. The single big chirp swelled and died and swelled, and as it moved through the thing it grew more and more complex, constantly modifying itself but returning to the same electrical pattern. The blob was *thinking*. It had somehow become an entity, not just a collection of two hundred million individual cells. And it had taken a bump from the outside world to make it recognize itself. Now there was clearly an outside world, and its own self. I analyzed its electrical commotion. It had no organized system of language, of course, but I could understand the electrical code and translate its meaning. Through a mist of confusion and primitive fragments of thought, a muffled message kept repeating: "Something is *out* there. Something is out *there*. Something is out there, and it has touched *me*." The "me" was the most beautiful part, a special electrical pattern created by many cells at once that could have no other meaning. Quite beyond any analysis of its individual cells, beyond its electrical and chemical impulses going this way and that, the thing had a sen-

sation of Unity. And, remarkably, I found that my feeling towards the thing had changed. Whereas before I
had regarded it as a mere mass of material, now I had
a sympathy towards the thing, even a tenderness. I
wanted to protect this little thing.

Two hundred million cells, more or less. And later,
cities, machines, symphonies.

Voices

I heard voices from the universe. From one star system, then another, then another. In the vast overlays of time and of space, inanimate matter was silent. But animate matter, matter with minds, spoke to me. Or more precisely, the creatures spoke to something they believed might be me. In some cases spoke to something they knew I was not. After all, I am what is, and I am what is not.

Blessed Mover, thank you for this feast. Thank you for what you have given me and my children.

What does it mean? This sky, this hand? Does anyone know what it means?

God will kill you for ruining my life. God, are you listening?

The beauty!

*I cannot bear the pain any longer. Please let
me die.*

*Curse you, Great Maker, or whatever you are.
I asked for rain, and you do not bring rain. You
are impotent. You are a fake.*

I'm late again. I wish the days could be longer.

*Is death the end? I cannot believe it is the end.
All of this. How can it end?*

"Such little lives," said Belhor. "Wouldn't you
agree? But there is also something of grandeur in
them. Not in the individual lives. The individuals are
just tiny specks, nothing. But in the monstrous jel-
lied masses of them, the crowds, the communes, and
planets, there is something of grandeur. They have
thoughts. And they strive."

"They strive for what they might attain," I said.
"And they also strive for what they cannot attain.
Most of them yearn for immortality. They want to live
forever, even though they do not know what forever
means."

Belhor and I were walking together through the
Void, just the two of us. He, as I, could hear the
voices. "Yes, it is strange," he said, "that they wish for
immortality. As far as they know, immortality could be
unending torture and excruciating pain."

"But they understand very well its opposite," I said. "They understand mortality."

"That they do," said Belhor. "They see death and dying all around them. They see other living things grow old, parents and loved ones. They see skin become brittle and dry. They see their ability to move slowly decrease, hearing and seeing diminish, internal organs fail one by one. Disease."

"You always describe things so grimly," I said. "Death is the way of all matter."

"It is your law," said Belhor. "It is what you wanted."

"I admire their dreams of immortality," I said. "It is noble to try to imagine the unattainable."

"I am not so sure it is nobility," said Belhor. "They don't want to die. It could be as simple as that. Many of them fear death."

"Yes, they fear. It is part of their suffering. I didn't want them to suffer. It grieves me that they suffer."

"Suffering is inevitable in living things," said Belhor. "And especially in creatures with minds."

"I have made suffering."

"You have created a universe with minds," said Belhor. "It is the nature of mortal minds to suffer, just as it is the nature of flesh to expire. The higher the intelligence, the greater the capacity for suffering. But you should not blame yourself for this suffering. They bring it upon themselves. It is not only their fear of death. They are laced with greed. And they desire to harm others. Even, ironically, to harm themselves. They kill. They murder. They wage war. They steal. They lie. Entire nations rot and waste away."

Let me lift myself up and devour my dead father, and then I'll be whole, rapturous.

Bring me courage and strength to kill that which wishes to kill me.

Eat. We are replenished.

Ah, what a beautiful body. The curve of my shoulders, my muscles. Yes. Yes. Thank you, God, for making me so handsome.

She is so young. Please do not let my child die.

We have found proof of the infinite. Here, look through this bright piece of glass.

What can I do? It is wrong to steal. I know it is wrong. But my mother has asked me to do it. We have nothing to eat.

"I have heard expressed two different views about mortality," said Belhor. "Some of the creatures believe that there is a new kind of existence after death. It is not a rational belief, but we have already acknowledged the limitations of the rational. Others believe that death is the final end."

"The fear of death in the second view I do not understand."

"Perhaps they get so enamored of life and the pleasures of life that they don't want it finished."

"That sentiment would lead to a great sadness, I should think, but not to fear."

"They might fear nothingness," said Belhor.

"What would you think about death, if you were one of these creatures?"

Belhor laughed. "But I am not one of these creatures. It is as impossible for me to imagine being one of them as it is for them to imagine being me. Or you."

"Many of them seem happy," I said, "glad to be alive and to be conscious of being alive. I see joy in the universe."

"More are arrogant and vain," said Belhor. "Consider the man admiring himself in the mirror. He reckons that he is superior because he has a beautiful body."

"I am not sure the fellow truly feels superior. If he is wise, then he knows that bodily beauty does not reflect inner value. His worship of his own beauty suggests that he does not believe that he has within himself anything else of value."

"But his wisdom is something of value. And if he is wise, he would recognize his wisdom."

"So we reach a contradiction," I said. "The man cannot be wise."

"Not for a moment did I think that this miserable creature is wise," said Belhor. "But he is not only unwise. He has contempt for those less handsome than himself. And for that, he deserves to be flayed and disfigured."

"You judge him harshly."

"He has brought this judgment upon himself," said Belhor. "I would deliver to him only what is proper. He, and others like him, demean what you have made. They make a mockery of your creation. But in fact, I would not bother. There are trillions like him. He is an atom. We cannot spend our time involved with individual atoms."

"Yet lives are lived by individual beings," I said. "Regardless of how many trillions there are, a life is an individual thing. Each life is precious."

"But of little consequence," said Belhor. "Certainly not for the whole. The case of the young woman whose mother has asked her to steal in order to support her family is more interesting. She is conflicted."

"I am concerned about her. She has already suffered a great deal with the death of her father."

"Do you think she would be justified in following her mother's instructions?"

"Why don't I look in more closely," I said.

"That is an excellent idea. We may learn from her case. Let us enter the new universe and observe."

"I will enter. But you should not."

"I have already done so," said Belhor. "Many times already."

"When? I did not give my permission."

Belhor smiled. He became thinner and thinner until he was a razor-sharp black line. The line stretched and stretched and extended through the vacuous space of the Void until it penetrated the pulsating sphere that was Aalam-104729 and disappeared.

"We are here now," said Belhor. He was hypnotic.

"Now that we are both here," I said, "you will respect my creation."

"Certainly," said Belhor.

"Well then, we will look in on the young woman, to learn from her life. I will unfold the folding of time, dissect the time of rotation of galaxies, then again and again, down to finer and finer durations of time, to individual lifetimes, to moments."

"Yes. You have captured a sliver of time, a slice of her life."

"In a certain cluster of galaxies. This galaxy. This star system, with three planets. This planet, the innermost planet. This dominion. This commune. This habitat. There. Can you see her? It is late afternoon, dusk. She gazes out of a window. Eighteen years old, in local years. Wearing a white mantle with a frayed border, she slumps against a stucco wall. There is another person in the chamber, her younger sister, who squats on the floor and drops little stones into a bottle one by one. A year ago, their father died an accidental death."

Every few moments, the young woman looks around the small chamber as if she were a visitor in her own abode, sighs, then turns again to stare out of the window. She gazes onto a small courtyard of spherical rocks and, beyond, narrow passageways barely wide enough for a cart, weeds hanging from the cracks in the stuccoed walls of other houses. Smell of rotting food. Her own habitat has a dirty tiled entry, a central chamber, two small sleeping spaces. On the pandamin, chimes sing in the breeze that comes before the rain. Other than the chimes, it is quiet, so quiet that the young woman can hear the tiny scratchings of an insect crawling across the limestone floor, and another soft murmur that is the sound of wind moving through the samarin groves.

The young woman has been standing by the window for some time, brooding in the dim silence, watching as the last rays of starlight stream through the west colonnade and cast long turquoise shadows across the floor and up the back wall onto the painted mural of an ocean storm. Now, her skull begins to pound. Whether it is anxiety or guilt, she doesn't know. She reaches up and gingerly massages her temples, then shrieks as another wave of pain surges through her head. At the cry, her younger sister leaps to her feet. "What is it?"

"Nothing," says the young woman. "There is meat in the latrum. Already cooked. Eat."

"Where did you get it?" asks the sister in astonishment. She crosses the room, then begins devouring pieces of the meat.

"Doesn't matter," says the young woman. "Save some for our mother."

"How did you get it?" asks the sister.

Without answering, the young woman walks across the chamber and ignites a copper lamp. It creates a distended ellipsoid of light and releases the fragrance of mlyex bark. Then she returns to the window. Outside, in the fading light, she sees a bedraggled man walk out of one of the cone-roofed dwellings, dump fish bones into the alley, and go back inside. A tadr bird circles the cistern three times before landing. Bad omens everywhere, she thinks to herself.

"What are you looking at?" asks the sister.

"Nothing," says the young woman. "I am not looking at anything."

"Then why are you staring out the window?"

"I don't know," says the young woman. "I did something bad today."

"Our mother will punish you. But I won't tell her.

What did you do? Was it really really bad? Does she know what you did?"

"She asked me to do it." The young woman continues to stare out the window and pushes her hand against a sharp edge of the window latch until it begins to bleed.

"What did you do that was bad? You don't have to tell me if you don't want to." The sister wipes her mouth. "Thank you for getting meat. Our mother will be happy. She said that she didn't know if we would ever be able to eat meat again." The girl looks down at the dusty limestone floor, shudders, and begins crying.

"Please don't cry," says the young woman. She puts her arms around her younger sister and kisses her forehead.

"What will happen to us?" says the sister.

"You should not worry about it," says the young woman. "I will get more good food."

"But what if you can't?" says the sister.

"You should not worry."

Immortality
Reconsidered

Uncle Deva and Aunt Penelope could not hear the voices. But I told them.

Poor girl, said Uncle Deva. She is so young. There is a great anxiety in her. And a sadness. In so many of these creatures. Uncle let out a long sigh. Now we know sadness. It has come into the universe, I don't understand from where, but I can feel it. Before now, I didn't know what sadness was. Now it has washed over houses and communes. It has infected lives like a wound that will not heal. I am so sorry, so sorry. And now I too am sad, for the first time in the infinity of time. That poor young woman, still a girl really. First her father died, and now this. What kind of mother would ask her child to steal?

The family needed food, I said. The mother hated to ask her daughter to steal, just as the daughter hated doing it. We had only a glimpse.

You cannot pay so much attention to individual

creatures in the new universe, said Aunt P. That is my advice. We must remember where we are and who we are.

Sometimes, Penelope, you sound so . . . I wish that . . . But I suppose you are right. Those beings are not of our essence. That is what I have been saying all along.

No, that is not quite what you have been saying all along, said Aunt P, and to punctuate her point she snipped off a piece of nothingness and stuck it in her hair. Before you were talking about inanimate matter. And then souls. Now you are lumping animate and inanimate matter together. Are they the same or not the same, which is it?

They are the same in some ways and not the same in others, said Uncle D. I am trying to agree with you.

Here we go, said Aunt P. What do you say, Nephew? You made all of it. It's your production.

I will not continue to be brought into these arguments, I said. Were you happier in that endless sleep we all had before time? When we were all doing a great deal of nothing? When there was nothing to do? It was easy, I admit, but is that what you wanted? I, for one, realize I was . . . *bored*.

Aunt looked away, as she did when she had been cornered.

I knew from experience that I had to allow her to save face. Animate and inanimate matter are made of the same material, I said. But there are obvious differences.

Of course there are obvious differences, said Uncle D. For one thing, the animate matter talks.

And apparently suffers anxiety and pain, said Aunt P. My dear husband, didn't you assure all of us that there would be no pain in the new universe? What about that rosy prediction of yours?

Uncle Deva was hurt by this remark and became quiet. He didn't deserve such treatment. You should not say such things, I said to my aunt. Please. You cannot criticize Uncle for being optimistic. The responsibility for suffering in Aalam-104729 is mine. I made the universe.

Aunt Penelope groaned and swept back her hair. Don't blame yourself, Nephew. None of us knew what would happen. The thing took off. You did a good job. It just took off by itself.

We looked over at Aalam-104729. The universe had grown so full and swollen that the pinch in its middle had all but disappeared. Since intelligent life had arisen, the sphere vibrated with more urgency, more intensity. Yet none of the dramatic happenings inside were visible from the outside. Outside, it still appeared as a pudgy sphere, silky and pink. As we were watching, it slowly glided by, tentative, always a bit separate from the other universes that zipped and slammed through the Void.

I don't blame your uncle either, said Aunt P. Sometimes things just happen, they just happen. She made an attempt at a little smile. I am out of sorts, out of sorts. I'm sorry, Deva. Forgive me, Deva. Then Aunt P did something I had rarely seen. She began weeping. None of us wanted the suffering, she said. It just happened. Or that terrible Belhor made it happen.

Uncle Deva embraced Aunt Penelope, and she allowed him to hold her. There, there, he said. The new universe is not all suffering and sadness. There is much happiness in the thing. Isn't there, Nephew. There is joy, and there is music, and there is spirit.

Yes, I said. All of those things. It is a beautiful universe.

We have been changed, said Aunt P. I can feel it. Everything is different now. The Void is different. What will become of us?

What do you mean by that? Uncle Deva exclaimed. We will go on, as we always have. We are immortal.

But I don't feel immortal, said Aunt P. Those poor creatures that you have made, Nephew, they will live and die. Why do we live on? Is it right that we should exist forever while they do not?

Penelope! What are you saying? said Uncle.

It just doesn't seem right, said Aunt P. Is it right that they will have pain, and we will never have pain? Those wretched creatures, and here we are strolling about in the infinite Void, infinite ourselves.

Uncle Deva was bewildered. Like me, he had rarely seen my aunt in tears. Penelope, didn't you say just a moment ago that we cannot pay attention to those creatures?

I don't know what I said a moment ago, said Aunt P. I am all turned around. The new universe has changed everything. Everything.

Yes, I said. As much as we want, we cannot help but feel for the creatures. Their lives. I . . . I never predicted.

You are definitely out of sorts, Uncle D said to Aunt Penelope. You are not acting yourself.

I don't know, said Aunt P. I just don't know. The suffering, the unhappiness. I don't know. Not right. What's happened? I just don't know. Suffering. Unhappiness.

We must think of the joy, I said. As Uncle has said.

But . . . that poor young woman, said Aunt Penelope. So many . . . so much . . . pain. And we . . .

Stop it! shouted Uncle D. You can't act like this. And stop that bawling.

Aunt P sniffled and shuddered and stood up. At full height, she was a rather imposing figure. Don't tell me how to act, she said. And don't condescend to me. Now, go fetch me my chair.

Certainly, my dear, said Uncle, smiling. Now you are acting like yourself. There's my old girl.

Like Diamonds

"Now that we've started," said Belhor, "we should see what becomes of the young woman. She is struggling to right herself like a turtle on its back. I trust you will not intervene?"

"It is not easy to watch her anguish," I said.

"You must allow her to make her own choices," said Belhor. "At the least, you must allow matters to take their course. You have already set everything in motion."

"I did not intend for this woman and her family to suffer," I said. "Your previous pronouncement, that suffering is warranted, is clearly not always true. This young woman did nothing to bring misery upon herself."

"But she made the decision to steal," said Belhor. "She could have decided otherwise."

"You know as well as I that there would be misery in either choice."

"Ah, but the choices are different, and the miseries different as well. She could have refused to steal. There are other ways she could have gotten food for her family. She could have begged for food, like her mother. Young and pretty as she is, she might have made a very successful beggar. Instead, she chose to steal. She made a decision. Not only did she steal, she stole from her neighbors, who knew her and trusted her. She violated their trust."

"She had no way of knowing whether she could get food by begging. That is *your* estimation, Belhor, not hers."

"And why does she weep?" asks Belhor. "Her weeping is not a matter of estimation. It is a fact. I believe she weeps out of selfishness. She regrets her decision and knows it will haunt her. She weeps from her guilt and anticipated suffering. She weeps for herself."

"I agree only in part," I said. "To me, her weeping shows that she is connected to the life around her, to her younger sister, to her mother, to her neighbors. She feels for them as well as herself. She is part of the world, shaking like a leaf pelted by rain."

Belhor and I had entered the universe again and now observed the young woman. She was walking along a narrow rock-strewn path between the houses in her commune, her head hidden within a shawl as if she did not want anyone to see her. It was early morning, and the cracked stuccoed walls glowed in the morning sun. The smell of her sweat mingled with the odors of cooking meat and smoke. Could I comfort her? She was so close. I watched her step after step. Could I help her? No, she was only one among many. I could not become involved. I could not. But there she was, so tender, in agony. Could I help her? Now? How could I watch without helping?

"I think you feel sorry for this girl who stole food from her neighbors," said Belhor. "What about the neighbors? Why not feel sorry for them?"

"I feel sorry for them as well. I could have intervened. I could have prevented all of this from happening."

"And would you also have intervened in the trillions upon trillions of other cases?" said Belhor. "After deciding which ones merited your intervention? And suppose you did intervene, with good intentions of course, but sometimes made bad situations worse? What then?"

"Stop taunting me, Belhor. I have said I will not intervene."

"And I am grateful for that declaration. Your non-intervention, in fact, is what makes these cases interesting. These cases have a certain . . . untidiness. But more than that, I maintain that your intelligent creatures must be able to make decisions on their own, without intervention, in order to know who they are. If they choose to do good, then they know something about themselves, and if they choose to do bad, then they know something else about themselves. Otherwise, they are like stones, they are inanimate matter. The creatures would be even more interesting if they were of any consequence. All little lives, such little lives. Still, from the aggregate we may learn something. And I find it amusing to see how they live— their cities, their habitats and rooms, their squalid little alleyways filled with garbage. Did you notice the dirty pools of water in the alley in front of the young woman's house?"

"Yes," I said. A thin film of pollen had floated down and covered their surfaces. The puddles of water split the sunlight and glimmered in colors. Like diamonds sprinkled about.

Hand Stunting

It has now been approximately 1.576 x 10^{33} ticks of the hydrogen clocks since I created the universe. Although I am the Creator, I have learned much from what I have created. One thing I have learned: the mind is its own place. Regardless of natural conditions and circumstances, even of biological imperatives, the mind can contrive its reality. The mind can make hot out of cold and cold out of hot, beauty from ugliness and ugliness from beauty. The mind makes its own rules.

Consider the case of the planet Uncle has named Akeba. It orbits the smaller of two stars in a double star system. Over the course of eons of evolution, the triumphant civilization on this planet has constructed a striking imbalance between its two genders. The females are considered to be inferior to the males. Not only inferior, but completely dependent. To ensure that the women will be helpless, the hands of all female children are rendered dysfunctional by severing certain nerves. After years of effective paralysis, hands shrivel up into little stumps of gnarled flesh. Females in this society cannot grasp objects, cannot engage in handcrafts or operate machines, cannot even feed themselves. Each female is thus completely reliant on males—that is, creatures with functional hands—to take care of her. For her entire life, she must be fed by males, she must live in the habitats built by males, she must be clothed and cared for by males. She must attach herself to a male and follow him all day. Female babies in this world continue to be born with normal hands, as millions of years of evolution have determined the survival benefit of such appendages, but the cultural traditions of this society oppose and contradict the natural. At an early age, the nerves are cut with a ritual knife. The secretions of the lstrex plant prevent pain.

One might dismiss such behavior if it occurred in a society of low intelligence, such as insects. But the inhabitants of this society on Akeba are mentally advanced and evolved in all other aspects. They (the males) build cities of ingenious design. They have created flying machines and electromagnetic communication devices. They celebrate their painters and musicians and writers and philosophers. Yet they accept their cultural tradition of hand stunting without questioning it. Males consider it both a duty and a pleasure to take care of the helpless females, oblivious to the fact that they, the males, have themselves produced the conditions in which females must be cared for.

The most unusual aspect of this hand-stunting tradition is that the females accept it without protest. Not that the females are stupid. They are every bit as intelligent as the males. But after many generations of the custom, the females, just as the males, take it as ordained. Mothers watch with approval as the ritual knife descends into the flesh of their young daughters and severs the critical nerves. To the mothers, this practice is as natural as rain. On the rare occasions that a woman challenges the custom, she is shunned by the community, given a dim-witted male to take care of her, and spends the rest of her life in loneliness and despair. The vast majority of females not only accept hand stunting but embrace it, seem content and even happy to completely surrender their independence to males. Their role in life is to be taken care of, to make the males feel powerful, and to mate with males for reproduction and pleasure. They welcome this role.

Despite my promise to Belhor that I would not intervene in the affairs of life-forms on Aalam-104729, I have been tempted to put a halt to the custom of hand stunting on Akeba. Do the females not wonder what they might accomplish with the hands they were born with? Do they not cringe from being treated as helpless pets, when they are as intelligent as their keepers? Do they not see that their dignity has been taken from them?

But I hesitate to intervene. For I cannot predict what would follow from my intervention. Perhaps without its time-honored tradition of hand stunting, the society would slowly disintegrate. Perhaps the creatures on Akeba have so adapted to the notion that males are superior, that females must be totally dependent on males, that they cannot live in any other way. Perhaps they cannot imagine a life in which females are equal to males. Their minds have created their reality. If both males and females are content with their roles, even happy with their roles, then would it not be best to allow the society to continue on its own, undisturbed, satisfied with its illusions?

If I do intervene, I would like to do an experiment. I would like to create a planet on which the custom is just the reverse: it will be the males who get their hands stunted. Then, after many generations in which this countertradition has been established, I would like to bring together the creatures from both societies, the one with helpless females and the one with helpless males, and see what happens. Each society should, I think, be shocked to see such a different reality and might finally become aware of its fabrications. But even with such knowledge, each society might prefer its illusions. The mind is its own place.

Religion

Both of you wanted the creatures in the new universe to have some awareness of me, I said. I wanted to tell you—

I thought you had already taken care of that item, said Aunt Penelope, fussing with her hair. Of late, Aunt P had been trying out new hairstyles and was at that moment inserting a fancy clasp made out of nothingness. Deva, she shouted, didn't you already sort that out with Nephew? The soul? The connection to Him?

I don't think He's done much about it, said Deva. I wouldn't—

I thought we had all of that sorted out, said Aunt P. You think something is sorted out. Then you find out it's not. If it's not one way, then it's the other. You think one thing, and it turns out to be the other. One thing, then the other.

You're babbling, dear, said Uncle D.

Well, Nephew, said Aunt P, what are you saying?

Did you take care of it or not? Do the creatures know that you are the Maker?

Yes and no, I said.

Here we go, said Aunt P. I tell you, the two of you are impossible, absolutely impossible.

The creatures have made up their own ideas about me, I said. They have *religions*.

What are you saying, Nephew? Did you straighten them out? Did you make an appearance?

An appearance? I said. A personal appearance! That would be way too much for them. And showy. I could never make a personal appearance.

So the creatures have *ideas*, without knowing anything about you for sure?

They have a lot of different ideas, I said. They want to believe in something big, to give meaning to their lives. They want some large purpose in the universe. I admire them for that.

I understand that perfectly, said Uncle Deva. These are intelligent creatures, Penelope. They think. They want meaning.

They want what *we* have, said Aunt P. They want immortality.

Of course they do, said Uncle. But since they know they can't have it, they want *something* to be immortal. They come and go so quickly. They want something to last.

But they don't know what they're talking about, said Aunt P. They are just guessing. I thought you were going to take care of this, Nephew. I thought you were going to let them know who you are, who you *really* are. Whatever they're imagining, it couldn't be what you really are. It couldn't be infinity. It couldn't be the Void.

From my observations, I said, I don't think they would be able to grasp the Void. They have no way of grasping it. But I have looked in on them. They have made beautiful buildings to worship me and to celebrate their belief in something eternal. I have seen the creatures congregate in those buildings, chanting and praying while waves of lavender and magenta light pour through arched windows or stream through openings in the domed roofs. They make offerings. They sing songs. They deny themselves comforts in order to live according to their beliefs. They teach their children their stories about the beginnings of the universe.

You don't have to say anything more, said my aunt. I know how you are. I thought you had taken care of this, but I can see . . . So, you are going to let them keep guessing.

Guessing is not so bad, I said. They feel a mystery about it all. I think a little bit of mystery is good. Mystery makes them wonder. It inspires them.

Mr g

Sometimes, Nephew, you are hopeless, said Aunt P. Well, you're going to do what you're going to do. So be it. They'll have their religions.

On a Small Planet

I did not tell Uncle and Aunt about all of my visits to Aalam-104729. Or of the many things that I saw. Once, I hovered invisibly in a city that arched over a hill. The planet was one of a dozen orbiting an ordinary star, the smallest planet in the system. It was a quiet world. Oceans and wind made scarcely a sound. People spoke to one another only in whispers. I floated above the city and looked down at its streets and inhabitants. Corners of buildings rusted in the air, billows of steam rose from underground canals. Through throngs of creatures moving this way and that, as creatures do in their cities, I spotted two men passing each other on a crowded walkway. Complete strangers. In the eight million beings living in the city, these two had never met before, never chanced to find themselves in the same place at the same time. A common enough occurence in a city of millions. And as these two strangers moved past, they greeted each other, just a simple greeting. A remark about the sun in the sky. One of them said something else to the other, they exchanged smiles, and then the moment was gone. What an extraordinary event! No one noticed but me. What an extraordinary event! Two men who had never seen each other before and would not likely see each other again. But their sincerity and sweetness, their sharing an instant in a fleeting life. It was almost as if a secret had passed between them. Was this some kind of love? I wanted to follow them, to touch them, to tell them of my happiness. I wanted to whisper to them: "This is it, this is it."

For Our Amusement

So much was new. So much was joyous. And disturbing. I asked Belhor: "Tell me, what do you think is the meaning of these creatures? What would you say is the meaning of their lives?"

Belhor laughed. "What meaning could they have? They amuse me. That is their meaning. But whether they have meaning in and of themselves? That would be giving these little things far more credit than they deserve. How can their lives have any meaning when they are mortal and know nothing of infinity? Their lives have meaning only insofar as they amuse us in the Void, only insofar as they increase our knowledge of what is possible and what is not. But meaning beyond that? No. What other meaning could they have?"

"I am not speaking necessarily of a grand meaning," I said, "but of individual meaning. Wouldn't you agree that each individual life has its own meaning, or at least a meaning as understood by the individual creature? Cannot each individual creature find some meaning for its own life?"

"What difference does it make?" Belhor said, and he stared as if bored at the gyrating spheres flying through the Void. On occasion, he would reach out and grasp one, squeeze it hard, then let it go. "There are so many trillions of 'individuals,' as you call them. They come and go. How could an individual mortal life have any meaning? And even if the individual, the tiny ant, *thinks* its life has a meaning, it is only an illusion. It is only a sensation, an excess of electrical current in its tiny brain. What significance could that have for us?"

"But surely it has significance for *them*," I said. "Each one of them tries so desperately to find meaning. In a way, it doesn't matter what particular meaning each of them finds. As long as each of the creatures finds *something* to give a coherence and harmony to the jumble of existence. Perhaps it might be as simple as a discovery of their own capacities, and a thriving in that discovery. And even if they are mortal, they are part of things. They are part of things larger than their universe, whether they know it or not. Wouldn't you agree?"

"With all due respect," said Belhor, "sometimes I cannot follow you. Why would you occupy yourself with such thoughts? How could you possibly suggest that these tiny ants have any consequence? Let them have their meanings. Let them have hundreds of meanings. If they like, they can believe that the cosmos is a giant fish, swimming in the mouth of a bigger fish. What difference does it make?"

Time Again

Time, always the magician. I now understand time to be my strangest invention. At moments slow, like the dripping of sap from a tree. At other moments quick, like the fluttering wings of a small bird. I go for brief expeditions in the Void, and when I have returned, eons have passed in the new universe. Civilizations flourish and fade during a single conversation with Uncle or Aunt. Or I can witness one event in the new universe, separate it into pieces of shorter and shorter duration, until shards of time are consumed with the wave of a hand, a breath, the first impulse of a nerve, the slight passage of one molecule. Eons to moments to eons.

Does time bring events into existence, or do events bring time into existence?

Not that time isn't exact. You can map events precisely onto the ticks of the hydrogen clocks. Assign each event a number, a precise position in the long fil-

ament of time. And yet what do you know in the end?
Uncle was right. You know little. You do not know the
cause of events. You do not know how events inter-
weave to produce new events. You do not know how
lived lives lie down, one after another, to make a verti-
cal pattern of action and change, or perhaps to form a
lengthy epoch of inaction and emptiness. You do not
know, from the temporal labels attached to a life, what
events brought happiness, or sadness, or joy, or regret.
Or simply boredom.

But the movements of electrons in hydrogen
atoms, the emission of photons with lockstep vibrations also do not know these things. They mark out
tick after tick after tick, in an unending stairway of
time, yet they know nothing. Of movements of gla-
ciers, of heartbeats, of gains and of losses, of individual thoughts they know nothing. Much less do they
understand.

Even I, who have infinite power. The unending
sequence of events, each meticulously labeled in
time, does not inevitably explain. There are things
that defy explanation: irrationalities, odd juxtaposi-
tions in time. For example: I do not yet understand
the life of the young woman who stole food for her
family. I have a complete record of every one of her
actions and thoughts. But I do not yet understand the
interplay of movements, the reasons for each event,

those that were accidental and those that were not. I do not yet understand which of her possible decisions would have been the *best* decision. That requires the future, but the future does not exist. Should she have disobeyed her mother, taken a chance that her family would starve, in order to uphold a principle of *right* behavior? Or should she have done as she did, violated her principles and beliefs in order to follow another principle: loyalty to her mother? Either way, she will almost certainly be haunted. Not all is logical. Perhaps even not all is understandable. After all, what is it that constitutes understanding? One can say that such and such an event came before another event in time. One can say that a system had a particular arrangement, which was a necessary consequence of a prior arrangement. Does such knowledge confer understanding? If events could be reshuffled in time, if future and past could be exchanged with each other, so that we always knew the consequences of actions before they occurred, would we then understand? Events—and the time that creates them or is created by them—cannot be contained. Events spill out and slide and defeat attempts to explain. Even against infinite power and force. This I have learned from the new universe.

And so, for some indefinite period of time, I am finished with this thought about time. Whether before something else, or after something else, or perhaps both, I am temporarily finished. The Void and the new universe, once so clearly distinct, are now part of the same fabric of time. Eons and moments. Mortality and immortality. That which lives and passes away, and that which lives on forever. Is it all not connected now? Existence and nonexistence? It is all connected now. I am all that is, and all that is not.

It is fair to say that the sadness Uncle Deva, Aunt Penelope, and I felt from the suffering of life in the new universe was more than offset by the evident joy of the same creatures. During our many visits to Aalam-104729, we witnessed happy celebrations of births, of marriages and unions, of natural events such as eclipses and solstices and air glowings, of symbiotic transferences, celebrations even of deaths. The two passing strangers I saw in the arched city. Despite their brief life spans, many creatures seem happy to wake up each morning, happy simply to breathe and to speak.

Nowhere is the joy of existence so apparent as in music. From one star system to the next, intelligent life-forms have created a multitude of sounds that express their exhilaration at being alive. There are waltzes and scherzos, apalas and calgias, symphonies, madrigals, fanbeis, sonatas and fugues, bhajans and dhrupads, tnagrs and falladias. The music dances and glides and swoops. Not that all of it is melodic or soft. But even the dissonant and the jarring contain a rapture, an ecstasy, an embrace of existence.

For some time now, I have admired many of the melodies invented in Aalam-104729 and find myself singing them as I move about in the Void. As do Uncle Deva and Aunt Penelope. We continue to sing our favorite songs trillions of atomic ticks after the composer has died, after the composer's civilization has vanished, sometimes even after the composer's central star has burned up and faded to a dead ember floating through space.

Before this new musical development, each of us often wandered alone through the Void, following one empty path after another during our excursions or simply searching for a solitary place to ruminate, shielded from one another by vast quantities of nothingness like beings on separate islands at sea. Out of sight, out of hearing, out of mind. We needed our privacies. Now, however, I no sooner set out on such a contemplative journey, glad to be alone among my own thoughts, when I hear Uncle or Aunt at some other location, loudly humming a tune picked up in the new universe. Will you please keep your singing to yourself, Aunt P shouts in the direction of Uncle D. I was having a pleasant stroll through the Void until you started up. You are one to talk, Uncle hollers back from a great distance away. I've been listening to that disagreeable song of yours now for eons, and I cannot hear myself think. Oh really, shouts Aunt P. Are you thinking or singing? Which is it?

When Aunt and Uncle get sufficiently annoyed with each other, they begin humming at high volume the very worst tunes in the universe, carefully selected from galaxy to galaxy and epoch to epoch.

Your voice sounds like the scaly underbelly of a bottom-feeding fish, Uncle D screams to Aunt P.

And you sound like a rotting pile of animal dung, Aunt replies.

Ha, ha, ha, Uncle shouts in a fake laugh. You know the difference between sounds and smells about as well as a rock can climb up a tree.

Dumb, dumb, dumb, shouts Aunt Penelope. I'm married to someone with an amber sunset for a mind.

All of us have taken to using metaphors from the new universe. Before, we had only the Void. And there are just so many things that can be compared to nothingness.

A Dress for
Aunt Penelope

There were other things we in the Void picked up from
the new universe. Birthdays, for example. We thought
birthdays were delightful. Aunt P and Uncle D began
planning birthday parties for themselves, although
choosing a date and an interval of time between
birthdays was a bit problematic. Following one of our
excursions into the new universe and back, Aunt P
suddenly announced that she was having a birthday
party after her next sleep, and she expected presents.
And I don't want any items from the Void, made out
of nothingness, she quipped. I've already got plenty
of nothing. I want something *material*. Do you understand what I'm saying?

A birthday party is a splendid idea, said Uncle D.
But about the presents, I don't know if—

No buts, said Aunt P. I'm going for my beauty rest
now. When I wake up, I want presents. Lots of them.

You've got a whole universe full of *stuff* to choose from. And I want something pink. With that, Aunt P yawned and retired.

A difficult spouse, Uncle whispered to me in exasperation. But what can you do?

So Uncle and I went into the universe and found a newly forming galaxy, full of pink stars, and we carried it back to the Void. In the Void, the material was nearly weightless. It shimmered and glowed. You could almost see through it. With a few folds and tucks, we made a beautiful dress for Aunt Penelope, as she'd always wanted. Uncle named the dress Kalyana. He said nothing about the mismatch of essences, as he had in mind a few material things he wanted for *his* birthday. Which, he informed me, was coming up quite soon.

Oh my! Aunt P exclaimed when she woke up and saw the dress draped over an outcropping of the Void. It is lovely. You shouldn't have gone to the trouble. Immediately, she put on the garment. Then she studied herself for a few moments. She turned this way and that, exclaimed again, and began dancing about and singing at the same time. Although there were many pink stars, there were also blue and yellow and green ones. It was a dress of many colors. Wearing it, Aunt Penelope was the most beautiful thing in the Void. Every time she twirled, a few stars came loose and began sailing off, and Uncle would shuffle over and scoop them up and stick them back on.

For eons, Aunt P never took that dress off, even for sleeping. It lasted for 10^{33} atomic ticks. After that, most of the stars had exploded or burned out, and the garment lost its color and shape.

Belhor & Co.
Go to the Opera

Uncle Deva, Aunt Penelope, and I were not the only ones making expeditions into the new universe. For some time, Belhor and his two assistants had been doing so as well, without my authorization. And while Belhor had beseeched me not to intervene in the course of mortal events, he himself had meddled from time to time.

On his first visit to Aalam-104729, Belhor landed at an opera house, on a planet almost completely covered with water. The intelligent life had constructed vast floating cities, and the opera house was located in the largest of these. As it happened, Belhor and the two Baphomets materialized during the middle of a performance, in the dark, suddenly occupying three vacant seats in the center front row. So as to attract as little notice as possible, the three visitors had taken the form of the local inhabitants. Belhor, tall and elegant, wore a charcoal black dinner jacket with tails, a starched white shirt with platinum cufflinks, a black bow tie, and shiny black shoes. Baphomet the Larger was dressed in a tan sports jacket several sizes too small, which strained to pop its single button at the waist, a wrinkled green shirt, polka-dot tie, and sandals. Baphomet the Smaller wore pajamas and slippers but had taken the trouble to put on a necktie, lopsidedly slung around his thick neck.

"The mezzo-soprano is off-key," said Baphomet L. after a few moments. "And I do not like her looks." The beast loudly leafed through the program notes until he found the singer's biography. "No, she will not do."

"We do not like her looks at all," said Baphomet S., who also riffled through the program.

"Quiet," said the patron sitting directly behind them. It was a woman who had spoken. She wore a taffeta evening gown with jewels in her hair, and she reeked of an unnatural scent.

Baphomet L. turned around in his seat and grinned at the woman.

"Surely, madam, you were not speaking to me," the beast said loudly, and rolled one of its two eyes clockwise and the other eye counterclockwise. At which point the woman gasped, leaped from her seat, and went hurrying up the aisle.

"Behave yourselves," whispered Belhor. "We are in a public place. We are here to observe."

"But the mezzo-soprano has an awful voice," complained Baphomet L. "We should speak to the manager and get our money back."

"We should demand a full refund," said Baphomet S.

"Quiet," said Baphomet L.

"Quiet both of you," whispered Belhor.

Almost imperceptibly, the theater rose and fell, as did all of the buildings in the floating city, and the overhead chandeliers slightly swayed.

"I am annoyed with that mezzo-soprano up there," said Baphomet the Larger. Suddenly, the mezzo-soprano, who was at that moment singing an impassioned aria with outstretched arms, discovered that the clasp holding her dress had come undone. The dress slithered down to her waist.

For the next few moments, the audience became very quiet. Then someone in the second balcony shouted an off-color remark, someone else began applauding, and the singer ran off the stage.

"Baphomet!" Belhor said sternly. "You disappoint me."

"But she had such an insufferable voice," said Baphomet L.

A young woman appeared on the stage and

announced with apologies that there would be a brief intermission. The lights went on. "I hope that they have got something good to eat," said Baphomet L.

"I'm famished," said Baphomet S.

A gentleman in a grey suit walked up to Belhor and tapped him on the shoulder. "The manager would like to see you, sir," said the gentleman. "Would you please follow me to the manager's office."

"And I would like to see the manager," said Baphomet L. "We want a refund. That mezzo-soprano was atrocious."

"What is the issue?" Belhor quietly asked the gentleman.

"Please come with me," said the man. He stared at Baphomet S., now standing in full view wearing his striped pajamas. "And bring your friends with you."

The three visitors followed the gentleman in the grey suit up the aisle through crowds of opera patrons, to a door at the far end of the lobby, up a stairway, and into the manager's office at the back of the mezzanine. The room was richly appointed with platinum fixtures, woven rugs, and furniture made out of crystal and glass. Photographs of the manager posing with various dignitaries covered the walls.

The manager was a middle-aged man with a soft and mushy face, like a piece of fruit past its prime, a limp moustache, and luminous green jewels on both

hands. For a moment, his gaze fell on the two Baphomets, one after the other. Then he addressed Belhor.
"May I see your ticket stubs?" Belhor produced three
ticket stubs. The manager examined them, then examined them again, as if something were not quite right.
"I am going to have to ask you not to speak during
the performance," said the manager. "You have been
disturbing other patrons."

"Certainly," said Belhor. "With all apologies. We
were . . . stimulated by the performance."

"I have not seen you here before," said the manager. "Is this your first time?"

"I have not seen *you* before either," said Belhor.
"How long have you been the manager of this theater?"

"If you were regular patrons," said the manager
with a slight smirk on his face, "you would know the
answer to that question." The manager looked again
at Baphomet S. and could not hide his disdain. "May
I ask your name?" he said to the beast.

"May I ask *your* name, my dear fellow?" said
Baphomet L. "You seem to be doing very well for
yourself."

"Very well indeed," said Baphomet S., gawking at
the glass furniture.

"You do not have appropriate dress," the manager
said. "Our ushers should not have let you into the
theater." The smaller Baphomet began crying fake

tears. "This is one of the most prestigious venues in the city," continued the manager. "Our patrons expect *refinement*. They deserve refinement. Royalty comes to our theater. Important guests come to our theater. Are you having fun with us?" At this point, Baphomet L. began hugging Baphomet S., pretending to comfort the beast. "Your friends should have more respect," the manager said to Belhor.

"You are insulting us," said Baphomet the Larger. "I do not like to be insulted."

"No, we do not like that one bit," sniffled Baphomet S.

The manager smiled. "You pretend to be something you are not," he said. "And then you mock what you cannot have."

"My gosh, you are right," Baphomet L. said, grinning, and he did a magnificent backflip. "You have found us out."

"I want all three of you out of my theater," said the manager. "Now. And don't leave through the main door. Mr. Thadr will show you out a back way."

"You have treated my friends discourteously," said Belhor. "I am not pleased with how you have treated them."

The manager laughed. "Take them out to the street," he said to Mr. Thadr. "And bring back an air freshener."

Suddenly, a woman flew into the manager's office, breathing heavily and appearing frightened. "I am . . . so sorry to interrupt you, Mr. Lzehr. I don't know . . . I don't know what happened. I had the money in the cash box, and it just . . . disappeared. It was there, and then it wasn't there. All the money from the night's ticket sales. It was there . . . and then it . . . was gone. I . . . don't know what happened."

"What a shame," said Baphomet L. "You have our sincere condolences."

"Our very most sincere condolences," said Baphomet S.

"We will take our leave now," said Belhor. "Please, Mr. Thadr."

In the confusion that followed, with ushers and cashiers rushing about, the manager barely heard the frantic message that a water pipe had just burst and the lobby was flooded.

Mind Planet

As eons went by, the civilizations in Aalam-104729 rose to greater and greater heights before they inevitably passed away and were replaced by new civilizations. I was continually surprised to see what new things the intelligent creatures would invent. They made machines that could fly. They built machines that could perform any mechanical task. They built devices to hear acoustic frequencies that their own sensory organs could not, see light that their own eyes could not, smell molecules that were undetectable by their own olfactory organs. They invented silicon-based electromagnetic communication devices that allowed them to see and talk to one another from great distances. They manufactured appendages and internal organs, so that they could replace defective parts in their own bodies. In some worlds, the creatures learned how to suspend their aging processes, so that they could take long trips between star systems and revive themselves upon arrival, many life-times later. On other planets, the creatures learned how to modify the replicating and control molecules so that they could create new life-forms that had never existed. Occasionally, the artificial organisms proceeded to infect and destroy all animate matter on the planet, but in other cases, the new life-forms could be engineered to cure diseases or provide energy or produce new types of machines, part animate and part inanimate. In some worlds, the creatures invented techniques to alter memories in their brains, so that they could have the sensation of experiences that had never, in fact, occurred.

The most fascinating invention—which occurred on a particularly advanced planet in a certain ellipsoidal galaxy—was the ability to separate mind almost entirely from matter. After eons of suffering from the various illnesses and disintegrations that necessarily accompany all animate matter, the creatures on this planet developed a method to extract the information in the cells of their brains and encode it in high-frequency electromagnetic radiation. These beings were completely devoid of bodies. They were pure energy. The only material substance remaining of each individual was a highly polished reflecting sphere, needed to contain and confine the mental radiations within.

To be sure, these creatures had to pay a heavy price for being thus encoded. They could no longer move about. They could no longer receive sensory impressions from the outside world, except for signals from other encoded brains. They could no longer see their thoughts carried out as actions. However, there were compensations. Once encoded, these creatures never suffered physical pain. They never had unsatisfied physical longings. They never grew hungry. They never grew thirsty. And they could live a very long time, their demise coming only by the slow degradation and information leakage of the electromagnetic radiation as it bounced back and forth trillions of times between the polished walls of the confining spheres.

These bodiless creatures truly lived in an interior world. It was a world of pure thought. It was a world of pure mind. Mountains, oceans, buildings could be imagined, but they existed only as abstractions, as slight alterations of electric and magnetic fields. Touch and vision and smell could be imagined, but they existed only as changes in frequency or amplitude of the energy waves that bounced back and forth, back and forth. Perhaps surprisingly, emotions could be felt. Long ago, the creatures, in their embodied form, had learned how anger, fear, joy, love, jealousy, and hate were imprinted in the biochemistry and electrical impulses of material brains. Thus, when the contents of their brain cells were analyzed and encoded in the delicate vibrations of electromagnetic waves, the emotions were transferred as well. Fear was a certain diagonal modulation of frequencies and amplitudes. Love was another. Jealousy was yet another. The bodiless creatures, existing entirely as energy waves, could experience all of these sentiments, if one interprets experience as the sensation of recognizing the meaning of certain repeating patterns of energy. But then, does that experience differ so much from the sensation of anger, hate, passion, and so forth in creatures with bodies, who recognize the meaning of certain repeating patterns of chemicals and electrified particles in their material brains?

Not only could these bodiless creatures experience emotions. They also had personalities. If a creature had been pompous and overbearing in its original, embodied form, then so it was in its bodiless, encoded form. If a creature had been meek and timid, fawning, good-natured, insecure, temperamental in its embodied form, then it would be the same when reincarnated as a concentration of pure energy. Likewise, if a creature had been a painter, a musician, a technician, a philosopher in its embodied form, it would be the same in its encoded form, since these talents and dexterities and intellectual capacities are, at their essence, properties of the mind. A painter, for example, could still create paintings, but the artistic aesthetic and design, the line and the form, would be translated to patterns of energy.

Furthermore, the painter or the musician or the philosopher could have personal relationships—friendships, business dealings, romantic liaisons—with other bodiless creatures, through the transmission and reception of radiation from one sphere to another. Although such relationships were strictly cerebral, they could be satisfying. If two creatures developed a romantic attachment to each other, they could share joy and pleasure and even a certain kind of sexual congress, all experienced in the exquisitely subtle transformations of their electromagnetic minds.

When one visits this planet, one beholds entire cities of these creatures. But instead of buildings, avenues, solar domes, bridges, one sees rows upon rows of little titanium spheres covering the hillsides and valleys. Love affairs, arguments, paintings, the discovery of scientific principles, even warfare are taking place within these spheres, yet are totally invisible from the outside. From the outside, one sees only rows upon rows of little spheres, motionless, soundless, serene. But I, who can see everything at once, know all that transpires within. I know that many of these bodiless creatures yearn for the bodies and physicality they once had. They are tormented. They worry that because their entire existence is now interior existence, then the exterior world might be only an illusion. Carrying this logic one step further, they worry that even their interior world might be illusion, that *all* is illusion. For how could they tell, within the confines of their little spheres, whether anything exists? All they know for sure is that they think. In a certain sense, isn't this true of creatures with bodies as well?

Good and Evil

"Please do me the honor to see where I live," Belhor said to me soon after Aunt Penelope's dress began to fade. "I think it will be worth your time."

"I know where you live," I responded.

"You know, but you do not know. Please. You will be my guest." As always, Belhor had a way of speaking that was hypnotic, a whispering voice that emerged from everywhere at once, like a wind that blows from every direction. And yet, as I have said before, no wind ever moved through the Void.

To get to Belhor's abode, we traveled an enormous distance. In fact, an infinite distance. But there are many orders of infinity, and after aleph-naught, we proceeded to aleph-vav, and then on to aleph-omega, and even on beyond that, to realms I had rarely inhabited. In a way that is difficult to characterize, the Void became thicker and thicker during our journey. Not that the Void has any substance or mass, but the layers of nothingness grew more compressed and dense, more tangled together, so that one had the sensation of moving through ever-thickening somethingness— like a gauze, to use a metaphor from the new universe. Moreover, as we came closer and closer to Belhor's domain, the music of the Void went through its own transmogrification, increasingly shaped and controlled by Belhor's thoughts as well my own. Dark fugues. Nocturnes. Symphonies of melancholia. It was all extremely beautiful but unsettling and sad at the same time, as if the Void itself were longing for something it desperately wanted. It also seemed that as we traveled farther and farther, we were descending, but that was only a sensation, since the Void has no down or up, no gravity. I can only report the impression of walking down a great stairway, down and down and down, farther and farther to some submerged depth in the emptiness. Billions and billions of descending steps. Even the ambient light of the Void—which, like the music, originates in my mind—grew dimmer and dimmer. We traveled in silence. Eons passed.

Eventually, we came to a magnificent floating castle. Its walls and surfaces, although made from layers of emptiness like all things in the Void, were so compacted and compressed that they seemed to have a materiality. Parapets carved with strange symbols thrust outward in peculiar directions. Towers and turrets shimmered in pale colors, first translucent, then utterly transparent, then translucent again. Through great arched windows, one could see courtyards and walkways, storerooms, great halls with ornate chandeliers, balconies and winding stairways, elliptical pools of liquid nothingness. Each of these structures emerged from the Void, vibrated and fluttered for brief moments, then dissolved into the surrounding vacuum. When the structures reappeared, they would be shaped differently or placed at different locations. At one moment, a round tower would appear next to a particular battlement. At another moment, both battlement and tower would melt away and reappear as a four-sided bell tower on the opposite wall. Each architectural feature of the castle was temporary and fleeting, of course, yet there was a density and momentary persistence I had seen nowhere else.

At the ramparts, I expected attendants and servants to greet us. When none appeared, Belhor explained that he lived in his castle alone. He seemed not unhappy with this solitary existence, and yet not altogether satisfied with it either—but far too proud to acknowledge a need for anything he did not have. For long swaths of time, he said, he never emerged from his castle. The two Baphomets evidently dwelled somewhere else, and he pointed vaguely to a dim region beyond the castle. "All of this," he said with a sweeping gesture, "as well as I myself came into being when you created the new universes."

"I did not knowingly make you," I said. Now that I had entered Belhor's abode, I felt an uneasiness that I was not accustomed to, almost an obligation to do his bidding, or at least to show sympathy to him even against my own judgment. As if he were slowly inhaling my independence and will. Was I his guest? Or an intellectual sparring partner? Or a target of his immense ego and force?

"No, you did not knowingly make me," Belhor said. "But you created certain *capacities*, let us say. When you created time and space and then matter, you created the *potential* for animate matter. And from there, it is only a few logical syllogisms to arrive at the existence of new minds. And then the ability to act, for good and for ill."

"And how would I destroy you?" I said.

"What an interesting question," said Belhor, "and straightforward. But I do not believe you would want to destroy me. You would have no one of my intelligence to converse with. To answer your question, it is not so easy to destroy me. You can destroy me only by destroying the worlds you have made. Is that something you wish to do? I do not think so. You are proud of what you have made, and you are rightfully proud." Belhor hesitated and stared at me, as he had often done in the past. "But why should we spend our time talking about destruction, when there is so much exquisite creation about, thanks to you."

During this conversation, we had walked through one great hall after another. In each hall, Belhor paused to adjust the trappings on one of his many thrones. Pictures of the Void adorned the shimmering ceilings, melting out of the nothingness, then slowly vanishing, then appearing again.

"I want to speak to you of good and evil," said my host as we walked through an interior courtyard. Leafy fronds of emptiness hung over pale green pools, bordered by flowing benches and couches and reclining chairs. Everything was immaculate. Everything shimmered into faint visibility, then dissolved, then reappeared in different form. "There are creatures in our new universe," said Belhor. "Creatures with great intelligence, who knowingly do harm and cause suffering. You have observed such things?"

"Yes," I said. "And it grieves me. All suffering grieves me."

"I admire your compassion," said Belhor. "But I want to suggest that evil, maliciousness, greed, deceit, while being unfortunate, are in fact *necessary*."

"Necessary? I could have prevented those despicable qualities. I should have prevented them." I felt that perhaps I should stop saying such things to Belhor. He seemed to take my sympathies for weakness. Belhor was a creature who understood only power and strength. Still, a most interesting creature. His mind.

"With all due respect," said Belhor, stopping to pick up a piece of fleeting debris on the mosaic floor, "I do not think you could have prevented those attributes once you brought the universe into existence. Those 'despicable qualities,' as you call them, are a mandatory part of existence, an unavoidable dimension of behavior."

"I cannot agree," I said. "I can conceive of creatures who are wholly good. If I can conceive of them, then they could exist."

"And what do you mean by 'good'?" asked Belhor.

"The doing of deeds to benefit others, for example. The living of a life devoted to beauty."

"Ah," said Belhor. "And please tell me, how do you know whether a particular deed *benefits* another being? And please tell me the definition of 'beauty.'"

"You understand these things as well as I."

"Yes," said Belhor. "But I understand them only because I understand their opposites. I understand evil. I understand ugliness. I maintain that good can be defined only in its contrast to evil, beauty only in its contrast to ugliness. Qualities such as these must exist in pairs. Good comes with evil. Beauty comes with ugliness, and so on. There are no absolutes in the universe, no unitary qualities. All qualities are bound to their opposites. Come, let me show something to you."

I followed Belhor through a corridor that led out to a coliseum, which was enclosed by a vast shimmering wall. The structure was decorated with a fine etching, intricate and beautiful. Upon closer inspection, I saw that the etching consisted, in fact, of names, billions upon billions of names. "Whenever a creature in the universe dies, its name appears on this wall," said Belhor. Indeed, during the few moments we lingered there, new names appeared so rapidly that the etchings proceeded to wind round and round on themselves, sprouting little shoots like the tendrils of a growing plant. "Listen," said Belhor. One could hear something like faint moans coming from the wall. And also faint laughter. But both so indistinct that it was difficult to tell one from the other. Moans and laughter, countless tiny voices, all crowding against and on top of one another, so that it seemed like a single sound, a great whispered rush of existence. "The recorded voices of the dead," said Belhor. "Recorded while living. So many little lives, each too faint to hear. And who would want to bother listening to one or the other of them. But added together, they make a perceptible sound. As you can hear, the individual sufferings and joys quite cancel themselves out. What remains is the great combined mass of them, a single breath."

"I know many of those names," I said.

"You astonish me," said Belhor.

"Your Wall of the Dead is fascinating," I said, "but it has little to do with our discussion."

"Yes, but I wanted you to see it. We were speaking of whether absolutes exist in the universe. I hold that they do not. There are only relativities. This you said yourself. In fact, didn't you decree that there would be no state of absolute stillness in the new universe? Wasn't that one of your organizational principles?"

I looked at Belhor without disguising my irritation. "I was thinking of physical principles, not principles of behavior or aesthetics. You know that."

Belhor laughed. "But did you not argue a while back that animate and inanimate matter should follow the same rules and principles?"

"Yes, I see your logic," I said. Belhor bowed. "But let us think about this for a moment," I continued. "I can agree that beauty and ugliness are relative concepts. A particular creature with six appendages on its body might be considered highly beautiful on its home planet and a horror on a planet where similar creatures had four appendages. But behavior is different, is it not? Isn't it true that a creature who kills another creature is committing a wrongful act, in absolute terms? I can answer my own question. If a creature kills another creature for personal survival, for food for example, this is not a wrongful act. But suppose a creature kills another creature when food is not the motive. Can we not say that this act is absolutely wrong, without reference to anything else?"

"What about the case of warfare?" said Belhor. "Suppose your creature is fighting in battle to defend its family, its commune. Under these conditions, is it not permissible, even honorable, to kill the enemy?"

"Yes. That is the same as for personal survival."

"What about to kill when one's honor has been disgraced?" said Belhor. "Or to kill in revenge when one's child has been murdered? How many more exceptions do you need before you will agree that absolute evil does not exist? Evil, good, beauty, ugliness—none of these can be determined in the absolute, without a particular context."

"Perhaps," I said. I found that a storm was now raging inside of me. I wanted to leave the castle, to walk in the open nothingness of the Void. I wanted empty space.

"If we agree that absolutes do not exist in any form," said Belhor, "then I believe that I have won my argument that good can exist only if evil also exists. Start with one quality in one situation, and it becomes the other in a different situation."

Long ago, perhaps eons ago, we had left the Wall of the Dead and were now moving through a palatial hall. Two grand thrones were placed at opposite ends of the hall, facing each other. "We have arrived here at last," said Belhor. "The Chamber of Triumph." Belhor looked contentedly about the vast hall. Then he took a seat on one of the thrones. "Both of us are more powerful than anything else in existence, are we not? Sit in the other throne, my friend, it is waiting for you. Sit."

"I cannot," I said.

"Isn't it a fabulous throne?" said Belhor. "Do you feel it is beneath you?"

"I will not sit here."

Belhor smiled and rose. "Suit yourself. If you will not sit on the throne, neither will I."

"We are not finished with our conversation," I said. "While you have made some valid points, I do not concede the argument. I do not believe that evil and ugliness are *necessary*, as you put it. Perhaps none of the qualities we have discussed are necessary. Perhaps we can dispense with all of these categories. Let the universe be as it is, without calling some things good and some bad, some beautiful and some ugly."

"That is an interesting point of view," said Belhor.

"But then, tell me why do some creatures feel a rapture when listening to music, or a thrill when seeing a wind move across a field, or a satisfaction when they have helped another being? We do not have to call such feelings good, or beautiful, but they exist nonetheless."

"Yes," said Belhor, "I grant you that. And some creatures gain pleasure from inflicting pain on others. And some spend their tiny lives living only for themselves. And some are crippled and wounded at a young age. We do not have to call these things evil or selfishness or suffering. But they exist. We can say that they are merely what is."

"I cannot accept these actions you mention even if they are what is," I said.

"Now you speak of acceptance," said Belhor. "That is another matter entirely. We must coexist with things we cannot accept, even when we have infinite power."

"Not all things can be contained," I said.

"Well spoken," said Belhor. "It is such a pleasure to converse with an intellectual equal."

"Is that an absolute pleasure or a relative pleasure?"

Belhor laughed, and the castle trembled with his laughter. "I will be very interested to see what becomes of your experiment with Aalam-104729. We will have to wait and see." Belhor gestured towards a curving staircase. "Come, we have not yet gone up to the towers."

Unlikely Companion

I thought. I did think. I am thinking. I will think.

For eons, it has been quiet and still in the Void. Aunt Penelope and Uncle Deva have been sleeping. It is almost the serenity of long long ago, before time. At that ancient moment of eternity, before time and space, all of my thoughts happened at once. But I did not have so much to think about.

Aunt Penelope and Uncle Deva have been sleeping. And I have been thinking. And while I have been thinking, the eternal music of the Void has softened and slowed like the breathing of a great animal falling asleep, with ever-increasing spaces between inhales and exhales, in and out, until there are eons between breaths, until each breath is a low moaning note.

I have been thinking about Belhor. A strange fellow, he is. As much as I detest elements of his character, he is the most interesting being I've encountered. At the beginning of the Era of Creation, I thought that perhaps I might converse with some of the more intelligent creatures in the new universe. And I do hear their voices. But they cannot hear mine. They can see what I've done, but they cannot hear me. Their intelligence is limited. Their understanding is limited, certainly. For how can they fathom the infinite? Or immortality? Or the Void? How can a creature of substance and mass fathom a thing without substance or mass? How can a creature who will certainly die have an understanding of things that will exist forever? All of these aspects of mortal existence have prevented a true communion between the new cosmos and me. But Belhor, like me, is not made of matter and flesh. Like me, he travels both in the material universe and in the Void, he can experience not only the one but also the many. Belhor, like me, can inhabit a realm without time and without space. Like me, Belhor is immortal. And he is immensely intelligent, even witty at times. He is my dim shadow. He is my antipodal companion. He is the thin black line. He is the beckoning voice in the Wall of the Dead.

I will wait and wait, until time has run out. And then I will exhale more time. But . . . not all things can be contained. That, I have learned. The suffering and joy, they cannot be contained, they spill, like events and like time. The irrational lives with the rational. Another thing I have learned, and this of myself: I can take chances. I can act, even with doubts.

I admit now, as I think with myself, that the "good"

and the "evil" are not easily defined—whether or not they are both *necessary*, as Belhor proclaims. Do not the circumstances themselves prescribe how one should act and behave? Who can decide in advance of the circumstances how one should behave? Of course, it helps to have principles that one can believe, but even these do not always determine what is good and what is bad. Perhaps the good is what makes a thing whole, makes a life harmonious with all its surroundings. Perhaps the good, like music, forms a completeness of being, while evil divides and fractures. I will wait and wait, until time has run out. And then I will exhale more time. But not all things can be contained. Even the questions cannot be contained.

The Void is nearly asleep. There I can see Aunt and Uncle, sleeping, sleeping, and they diminish to two dancing dots, dancing a waltz that moves slower and slower. How long will they sleep? Eons. And Belhor sleeps in his castle. But the new universe does not sleep. It unwinds and evolves, it builds and destructs, it sings and it sings and it spins to the future.

Uncle Deva's Dream

Uncle Deva has awakened. In an unusual break with past habit, he has bounded up with ideas and energy while Aunt Penelope remains asleep. Ever since her birthday party, she has slept longer and longer intervals and, when awake, spends more time brushing her hair.

I had a dream, said Uncle D, stretching and yawning. It was not a very restful sleep. I dreamed I was taking a long walk through the Void, in places I had not ventured, and then there they were, thousands of creatures from the new universe, all begging me for second lives. Of course, I referred them to you, Nephew. I didn't know what to say to them. And you told them that they had to go back inside the universe and be dead. They didn't take the news well.

An unpleasant dream, I said to Uncle. I can see that you are distressed.

Yes. The dream quite woke me up. Also, your aunt

snores more loudly than the sharpened spines of a puya. Nobody can sleep next to her.

You don't have to worry, I said. Material life-forms could never get out of the universe and into the Void.

I understand that, said Uncle D. But they looked so forlorn . . . Nephew, now that the thing has been running for a few zillion ticks, I don't know how many, couldn't we let the creatures have a little bit of a second life, at least the intelligent ones? Some little remnant of them that continued on? Their lives are so quick.

But they are material, I said. It would have to be a nonmaterial remnant, and then it would be different from everything else in Aalam-104729. You've said yourself that the things in the new universe have a different essence than things in the Void. It is the nature of material things to pass away.

I know, I know, said Uncle D, and he sighed. But there must be something. Perhaps just at the moment of death they could *feel* a little piece of the Void.

Some of them have that feeling now, I said. At least they feel a mystery. They sense that there are big things they do not know, even though they have no way to know them.

That's what you said before, said Uncle. Religion. But a mystery is mysterious. I want them to have something more than that.

Mr g

At that moment, we heard Aunt P rustling about, calling for her slippers. She's up, said Uncle. But she can wait. Nephew . . . those figures from my dream haunt me. I can see them still. One mortal life isn't enough. Cannot we do something for them, something that will not violate your precious rules of materiality? Uncle reached over and plucked up Aalam-104729, which had been slowly drifting by. Although it was continually expanding, it did not seem as swollen as before, as if it had successfully digested a big meal. So much inside this thing, said Uncle. And yet I can hold it so easily. So many creatures yearning for more life, yearning for something more.

Perhaps we could give them just a tiny glimpse of the Void, I said. But I worry that it might be too much for their minds.

Please, said Uncle. Try. I would be so relieved if the intelligent ones could understand that their lives, as brief as they are, are part of something immortal.

Yes, I said, I believe I can do that.

At the moment of death, said Uncle. Then, even if the glimpse is overwhelming, it will not crush their lives.

Let us try with the guilt-ridden young woman who stole food to support her family, I said. Eons ago. I will shift time, so that we can go back to her era. And . . . I have wanted to do something for her. This, at least, will be something. Even though at the end of her life.

Thank you, Nephew. We will not be gone long? Your aunt is getting impatient.

Then I compactified Uncle to a dot, and we entered the universe. Together, we glided through the cosmos, through galactic clusters, past one shining galaxy after another, to a particular galaxy, one spiral arm, one star system, one planet, one commune.

There she is, now eighty-three local years old. She lies upon a porcelain bed in a dim chamber, surrounded by her children and grandchildren, the scent of the ritual branda plant filling the air. Following her custom, she lies on her right side, one of her eyes covered with a leaf. Her labored breaths make a sputtering sound. You hear her breathing? Yes, says Uncle, the time must be near. It goes by so quickly, a life.

Sixty-five years have passed since she stood by a window in a different abode in a different commune, looking out at a courtyard with spherical rocks and a tadr bird circling the cistern. Soon after that defining moment, struggling with confusion and guilt, she left her home, joined a traveling group of merchants, decided to punish herself by uniting with a vagabond, gave birth to a child that she abandoned. Every man she met, she wanted to be her father. Over time, a great remorse flowed through her, and she forgave her mother and herself. But she did not forgive the vagaries of the universe. She and her second husband had four children, who then bore their own children. Never did she find the child she abandoned in her youth. Despite years of searching, she never found that child. And that is her final regret as she now lies on her right side, breathing with shallow breaths, holding the hand of her grandson.

She has had moments of joy in her life, as well as frustration and sadness? asked Uncle.

Yes.

Sixty-five years passed in an instant. Her life has been used up in an instant.

Her ability to hear has ceased. Although one of her eyes is open, she sees only dim hazy shapes. She feels heavy, as if she cannot move from the position where she lies. Her mouth feels dry. She cannot move her tongue. Her breathing sounds like a pant, a dry thirsty exhale and inhale.

The moment approaches. She dreams.

Through the opaline clouds of a dream, she sees herself as a girl. Her father, young and strong, runs with her over a field. He is trying to tell her something, but each time he speaks she cannot hear him. Then she is holding her first child, the one she abandoned, the one she loved most. Then she is walking from habitat to habitat making her prophecies, behind each door a face, a room, a table, running water.

I can't believe it's over. I don't want to see it. I don't want to hear it. It grows smaller and smaller. Where's my daughter Leita? Where's my son Mrand? I want a second life. I want more life. Please. You, there. Who are you? What is this place? Is this where I've gone?

She has entered my dream, said Uncle Deva. Uncle began weeping.

Now, I said. Now she has the glimpse.

The old woman lying on her right side gave one long exhale, and a smile appeared on her face, and she died.

Permanence from Impermanence

And she died. At that moment, there were 3,147,740, 103,497,276,498,750,208,327 atoms in her body. Of her total mass, 63.7 percent was oxygen, 21.0 percent carbon, 10.1 percent hydrogen, 2.6 percent nitrogen, 1.4 percent calcium, 1.1 percent phosphorous, plus a smattering of the ninety-odd other chemical elements created in stars.

In the cremation, her water evaporated. Her carbon and nitrogen combined with oxygen to make gaseous carbon dioxide and nitrogen dioxide, which floated skyward and mingled with the air. Most of her calcium and phosphorous baked into a reddish brown residue and scattered in soil and in wind.

Released from their temporary confinement, her atoms slowly spread out and diffused through the atmosphere. In sixty days' time, they could be found in every handful of air on the planet. In one hundred days, some of her atoms, the vaporous water, had condensed into liquid and returned to the surface as rain, to be drunk and ingested by animals and plants. Some of her atoms were absorbed by light-utilizing organisms and transformed into tissues and tubules and leaves. Some were breathed in by oxygen creatures, incorporated into organs and bone.

Pregnant women ate animals and plants made of her atoms. A year later, babies contained some of her atoms. Not that her atoms had identification labels. But they were certainly *her* atoms, there is no doubt about that. I knew which ones. I could count them. Here, and here, and here.

Several years after her death, millions of children contained some of her atoms. And their children would contain some of her atoms as well. Their minds contained part of her mind.

Will these millions of children, for generations upon future generations, know that some of their atoms cycled through this woman? It is not likely. Will they feel what she felt in her life, will their memories have flickering strokes of her memories, will they recall that moment long ago when she stood by the window, guilt ridden and confused, and watched as the tadr bird circled the cistern? No, it is not possible. Will they have some faint sense of her glimpse of the Void? No, it is not possible. It is not possible.

But I will let them have their own brief glimpse of the Void, just at the moment they pass from living to dead, from animate to inanimate, from consciousness to that which has no consciousness. For a moment, they will understand infinity.

And the individual atoms, cycled through her body and then cycled through wind and water and soil, cycled through generations and generations of living creatures and minds, will repeat and connect and make a whole out of parts. Although without memory, they make a memory. Although impermanent, they make a permanence. Although scattered, they make a totality.

See, Uncle, it is done.

Material
Intelligence

Eons passed in the universe. But in the Void, eons can be moments.

"The Void does not seem what it was once before," said Belhor. "Would you agree?"

"It seems much more empty," I said.

Belhor laughed. "Isn't it fascinating that a totally empty thing can become more empty. A long time ago, I predicted that our new universe would change us. Indeed. We have all become more full, and everything else has become more empty. But the new universe is not so new anymore, is it."

"It is nearly 2.5×10^{33} atomic ticks old."

"It is passing away," said Belhor. "Already many of the stars have faded. Even a universe passes away."

"I say good riddance," said Baphomet the Larger, who was following behind Belhor and me as we moved through the Void. "The place had some unpleasant individuals in it."

"Some *very* unpleasant individuals," said Baphomet the Smaller, who walked another few paces behind Baphomet the Larger. "But we fixed quite a few of them, didn't we. We fixed them good."

Baphomet L. turned around and frowned at Baphomet S., then did a backwards somersault. At which point the smaller Baphomet performed his own somersault, clumsily done. The larger Baphomet stopped and showed the smaller beast how to tuck itself in during the roll.

"None of you should have intervened," I said. "I did not intervene."

"Allow me to apologize for the enthusiasm of my companions," said Belhor. "But we did not alter the course of events in any significant way. We merely observed. And if we ever did more than observe, it was only to give a slight nudge to capacities and tendencies already there, to events already in motion."

"You have a circuitous way of saying things," I said.

"You are getting to know me," said Belhor. "Nonetheless, I think we can all agree that the thing had quite an inertia of its own. The universe and its contents, including its minds, seemed to know from the beginning where it was headed."

"The most advanced civilizations continually amazed me," I said. "On their own, they discovered my organizational principles. With not much to go on. I have been impressed."

"Yes," said Belhor. "Impressive. But none of them ever discovered the First Cause."

"No, that would have been impossible. With their calculations, they could go far back in time, back to the point when the cosmos was only a fraction of a tick old. But there is no way they could go back to the beginning."

"Of course not," said Belhor. "They cannot get outside of the sphere they inhabit. They cannot even see the walls of the sphere. For all of their inventiveness, they are still insects compared to us."

"They exist in three dimensions. But it would be the same at five or a hundred. As you say, they cannot get out of the space they inhabit. They cannot even imagine the Void."

"I, for one, would not be comfortable with their imagining the Void," said Belhor. "Let them live and expire inside their little sphere. All in all, it has been an interesting experiment."

"Experiment? Experiment?" said Baphomet the Larger. "Nobody told me we were doing an experiment."

"Nobody told me either," said Baphomet the Smaller, who pretended to start sobbing. "Nobody ever tells me anything."

"We have learned something about what a material mind is capable of," said Belhor. "It is capable of great goodness, and also great evil. And more extreme in either case than I would have thought. But that is what happens with intelligence."

"Is it a consequence of intelligence, or materiality?" I said. "Because *we* have even greater intelligence."

Belhor smiled and said nothing.

Nihāya

And the universe continued to age and grow old. One by one, the stars burned up their nuclear fuel. Without replenishment, their internal heat slowly leaked away. And without heat and pressure to support them against the inward pull of their own gravity, the spent stars contracted and shrank and dwindled until they were cold and dim embers floating in space.

Planets in orbit about these dead stars no longer had a ready source of energy. Consequently, their life-forms could no longer sustain themselves. Here and there, some few civilizations had created their own energy sources, independent of their central stars, but all energy is limited, and in time these sources too were depleted. In eons past, after the first generation of stars had passed away, a second generation had formed from the contraction of great clouds of gas. Now, however, the universe had expanded and spun out to such a degree that the gas filling the vast spaces between stars was too thin and sparse, and it was completely unable to form new stars. Gravity, once the creator of stars and the creator of life, was at this point only a feeble force in the universe.

Slowly, slowly, animate matter became extinct. The tiny fraction of material in the cosmos that had been alive, the smattering of mass that existed in the form of living and breathing beings, diminished to zero. Once again, Aalam-104729 was a sphere of dead, life-less matter. Only now, all of its energy was in unus-able form. The potential for life had been utilized and exhausted. The only future for Aalam-104729 was to continue expanding, dimming, thinning, with the par-ticles of its dead matter getting farther and farther away from one another.

There came a point in time when only a single gal-axy remained that harbored life. Uncle Deva named this galaxy Nihāya. Stars in all other galaxies had faded away, but in this galaxy, a few points of light still shined. There came a point in time when only a single planet in this galaxy contained life. The intelligent beings on this planet understood that the universe was dying, that their days as a civilization were num-bered. But they could not know that they were the last in the universe. They made paintings and music and books to commemorate the end of their existence, but they could not know that there would be no future beings to witness these things. The last life on this planet was, in fact, insects and plants. There came a point in time when this too passed away. And life in the universe was finished.

"Do you have regrets?" said Belhor to me. "Knowing that there would be suffering, would you make the universe again?"

"I regret that there was suffering," I said.

"Yes, but would you do it again?"

I entered Aalam-104729, and I glided past the dim galaxies, the dim planets, the dim stars. Again I heard voices. Voices from past civilizations that dreamed of immortality.

A wise man, recognizing that the world is but an illusion, does not act as if it is real, so he escapes the suffering.

We have built the heaven with might, and We it is who make the vast extent.

The senses are higher than the body; the mind higher than the senses; above the mind is the intellect; and above the intellect is the Self.

What is seen is temporary, but what is unseen is eternal.

A New Dress for Aunt Penelope

From the outside, Aalam-104729 looked as it always had. Aunt Penelope held it up, shook it, and listened. Then she sighed. Well, Nephew, she said, what will you do for your next one? As I remember, you wanted everything bigger.

I changed my mind, I said. I'd like the next one to be the same.

As we were speaking, billions of empty universes flew through the Void, all waiting, and waiting. They hummed and sputtered and shrieked.

The next one won't be the same, said Uncle Deva.

No, I said, it won't. But I would not be unhappy if it was very similar.

I think we were all rather fond of the thing, said Uncle D.

Yes we were, I said. And fond of the inhabitants.

It was a lovely thing, said Aunt P.

It was a beautiful thing, said Uncle. It had beauty. And joy. And sadness.

It had all of that, I said. It had everything.

Not everything, said Aunt. It did not have immortality.

No it did not, said Uncle. But I think maybe it did have a soul after all.

A *mortal* soul? said Aunt. Sometimes I can't tell what you are talking about, Deva.

It's gone now, I said. That is the nature of things. But I do miss it.

There will be another one, said Deva. And another one after that.

Yes.

OK, Nephew, said Aunt P. I'm anxious to see what the next one will be like. Let's get on with it.

Don't rush Him, said Uncle. He needs to take His time.

Yes yes, said Aunt. And just so that both of you know, I'd like a new dress from the next universe, like the last one.

Certainly my dear, said Uncle.

Notes

Origin of Names

Aalam is a Muslim name meaning "universe." Belhor, also called Belial, Baalial, and Beliar, is a demon figure in the Christian and Hebrew apocrypha. Baphomet is a twelfth-century pagan deity in Christian folklore, appearing in the nineteenth century as a Satan-like figure. *Deva* is a Sanskrit word meaning "deity." *Nihāya* is an Arabic word meaning "ending," used to refer to the endings of Arabic stories and Sufi poetry. Ma'or is a Hebrew name meaning "star." Al-Maisan is an Arabic name meaning "the shining one."

Science

The physical creation of matter and energy, galaxies, stars, and planets, and the emergence of life follow the best current data and theories in physics, astronomy, and biology. All quantitative discussion of various cosmic events is scientifically accurate. The unit of time used by Mr g, the "tick of a hydrogen clock," is the reciprocal of the frequency of the Lyman alpha emission from the hydrogen atom, equal to about 4×10^{-16} seconds.

Numbers

There is no reason why Mr g would use base 10 to discuss numbers, but this base has been used here because it is the number system that will be familiar to most (Earthling) readers.

Origin of Italicized Quotations
at End of Chapter Nihāya

"A wise man, recognizing that the world is but an illusion . . ."   —saying of the Buddha

"We have built the heaven with might . . ."
                                    —Qur'an, 51:47

"The senses are higher than the body . . ."
                                —Bhagavad Gita, 3:42

"What is seen is temporary . . ."
                    —New Testament, 2 Corinthians 4:18

Alan Lightman is the author of five previous novels, a book-length narrative poem, two collections of essays, and several books on science. His work has appeared in *The Atlantic, Granta, The New Yorker, The New York Review of Books*, and *Nature*, among many other publications. His novel *Einstein's Dreams* was an international best seller and has been translated into thirty languages. His novel *The Diagnosis* was a finalist for the National Book Award. A theoretical physicist as well as a novelist, Lightman has served on the faculties of Harvard and MIT, and was the first person to receive a dual faculty appointment at MIT in science and in the humanities. He is also the founding director of the Harpswell Foundation, which works to empower a new generation of women leaders in Cambodia. Lightman lives in the Boston area.